Vibrations of Power Plant Machines

Franz Herz · Rainer Nordmann

Vibrations of Power Plant Machines

A Guide for Recognition of Problems and Troubleshooting

 Springer

Franz Herz
Energy Consulting Group
Wohlenschwil, Switzerland

Rainer Nordmann
Fraunhofer Institute for Structural Durability
and System Reliability LBF
Darmstadt, Germany

ISBN 978-3-030-37346-7   ISBN 978-3-030-37344-3   (eBook)
https://doi.org/10.1007/978-3-030-37344-3

© Springer Nature Switzerland AG 2020
This work is subject to copyright. All rights are reserved by the Publisher, whether the whole or part of the material is concerned, specifically the rights of translation, reprinting, reuse of illustrations, recitation, broadcasting, reproduction on microfilms or in any other physical way, and transmission or information storage and retrieval, electronic adaptation, computer software, or by similar or dissimilar methodology now known or hereafter developed.
The use of general descriptive names, registered names, trademarks, service marks, etc. in this publication does not imply, even in the absence of a specific statement, that such names are exempt from the relevant protective laws and regulations and therefore free for general use.
The publisher, the authors and the editors are safe to assume that the advice and information in this book are believed to be true and accurate at the date of publication. Neither the publisher nor the authors or the editors give a warranty, expressed or implied, with respect to the material contained herein or for any errors or omissions that may have been made. The publisher remains neutral with regard to jurisdictional claims in published maps and institutional affiliations.

This Springer imprint is published by the registered company Springer Nature Switzerland AG
The registered company address is: Gewerbestrasse 11, 6330 Cham, Switzerland

I dedicate this book to my sons Harald, Johann and Stefan.
—Franz Herz

Preface

Coming to a power plant as an expert, because of a vibration problem, you will hear in the most cases from the maintenance people: When do we have to shut down for balancing? Rebalancing seems to be the only generally known countermeasure for high vibrations at a turbomachine. This book shows that a mechanical unbalance covers only a small section in the big line of all the possible vibration problems.

The aim of the book is to provide practical people working in the field, at power plants and/or as service engineers, with a guideline in case there are vibration problems. In the books available on vibration, there are usually a lot of theories, equations and high mathematics. This book does not make great demands on scientific perfection, it is meant for the practical "hands on" man.

Many of the illustrations and sketches in this book come from old reports that are no longer available. Therefore, the quality of the illustrations does not always meet today's standards.

However, I have decided to include these images in this book, because they are valuable, given that they are the only images of their kind that still exist.

In Chap. 1—"Basics of Vibrations"—some fundamental theories are explained, which is necessary to understand the vibration events mentioned in Chap. 3 —"Fault Analysis: Vibration Causes and Case Studies."

This book does not have the usual references as it includes mainly real-life case studies originating from the 50-year experience about power plant events. These real-life case studies could serve as a reference, regarding the possible problems at jobsite.

It is tried to describe the case studies in three steps, explaining the:

1. identification of the problem,
2. explanation of the problem and
3. practical solution to the problem.

Also, some knowledge about measurement and presentation of results is not avoidable; therefore, see Chap. 2—"Instrumentation, Measurement."

The rotors of a machine are balanced individually by the manufacturer. They will be coupled to a shaft train at first at jobsite. So, a new unbalance distribution is created and might require a balance correction. On top of that, the newly installed rotors will see operational parameters (like temperature, torque, etc.) the first time. This might be another reason for a necessary balance correction. Therefore, Chap. 4—"Jobsite Balancing"—has been added. It is not meant as a general balancing instruction, but as a guideline for trim balance corrections for the above-mentioned occurrences.

There are a lot of books on vibration on the market, but not in one of them we could find a comprehensive collection of case studies of real cases. This book had been written in English, because according to our experience, especially at the East Asian region, there seems to be a need for comprehensive practical machinery vibration primer.

Wohlenschwil, Switzerland   Franz Herz

Darmstadt, Germany   Rainer Nordmann
                                                nordmann@rainer-nordmann.de

Contents

Chapter 1
Basics of Vibrations

This chapter mainly deals with the relationship of vibration excitation and vibration response of systems. Firstly, we analyze the different vibration signal types and the different vibration measurements units. This leads us to the Fast Fourier Transformation (FFT). We now can explain the resonance frequency: $\omega_0 = \sqrt{\frac{c}{m}}$. The vibrations of a single degree of freedom (SDOF) are explained then and finally the rotating shaft is explained by means of the Laval shaft. The last chapter deals with the practical behavior of turbomachine rotors.

In Chap. 3: "Fault Analysis: Vibration Causes and Case Studies"—there are various vibration problems specified, as they may appear in power plant machines. To understand and interpret these vibration phenomena, we need to understand the basic of vibrations.

At first, we must ask how we can describe vibrations of mechanical systems in terms of deflections, velocities and accelerations as a function of time. This consideration of the kinematic of vibrations is independent from the vibration system and from the source of vibrations. Important quantities to describe, for example, a simple harmonic vibration are the time period and the frequency of the vibration (see Sect. 1.1.1). However, very often vibrations cannot be described by one frequency only, but they consist of a superposition of time signals with different frequencies. In order to recognize these different frequencies involved in a vibration event, the Fourier analysis is a powerful tool (see Sect. 1.1.2). The important relations between vibration deflections, vibration velocities and vibration accelerations are derived in Sect. 1.1.3 for the simple case of a harmonic time signal, expressed by amplitude, frequency and phase.

The second important question is: What causes the vibrations of mechanical systems? The theory of mechanical vibrations shows that vibrations depend on some kind of excitation on one side and the dynamic characteristics of the vibrating mechanical system itself on the other side. Excitations can be time-dependent forces and/or moments, but movements of the ground or other boundaries are also possible. Excitations may be different in the time domain, where the time functions can be harmonic, periodic or non-periodic (transient). An excitation can be of very short time or may act permanently.

© Springer Nature Switzerland AG 2020
F. Herz and R. Nordmann, *Vibrations of Power Plant Machines*,
https://doi.org/10.1007/978-3-030-37344-3_1

The other important influences on vibrations are the dynamic characteristics of the mechanical system itself. The physical system parameters of mass, damping and stiffness values determine how a vibration system reacts to excitations (disturbances). The dynamic characteristics of a mechanical system can also be expressed by its eigenvalues (natural frequencies, damping) and mode shapes or by frequency response functions (FRF). From a more practical view, mechanical systems react very sensitive with respect to vibrations, when they are excited in a resonance condition (exciter frequency equal to a natural frequency). In the case of rotating machinery, such resonance conditions are called critical speeds, where the rotational shaft frequency of an unbalance excitation is equal to one of the natural frequencies of the mechanical system.

Therefore, at each vibration problem two fundamental aspects must be considered:

1. the excitation forces like the unbalance forces due to rotation of the shaft and
2. the consequence of those forces to mechanical systems upon which they are acting like in critical speeds and in resonances.

The engineer in charge must decide what is the most promising way to overcome a problem: Is it 1 or 2 or perhaps both.

A very simple mechanical system to explain the basics of vibrations is the single degree of freedom (SDOF) system (see Sect. 1.2), consisting of the two parameters mass m and spring c. For this basic system, the equations of motion, the natural frequency, the free vibrations and the forced vibrations in case of a ground excitation are derived and discussed. The additional effect of damping on free vibrations (see Sect. 1.2.1) and forced vibrations (see Sect. 1.2.2) are investigated. From the simple mass-spring system, we lead to the rotating shaft since it also follows similar fundamental laws.

The transfer from the above SDOF mechanical system to a similar system with a flexible rotating shaft is shown in Sect. 1.3. This basic mechanical system is the Laval rotor, where the mass m is represented by a disc in the center of the shaft and the spring c by the bending stiffness of the rotating shaft. Extensions of this very basic system are possible, e.g., by the flexibility of the bearings and the support system, which may include additional springs and masses as well. The effects of rotation, e.g., unbalance and gyroscopic, must be considered.

The vibration system of a complete power plant machine, e.g., a turbine shaft train consisting of the shaft train, the oil film bearings, the pedestals and the foundation, is much more complicated and must be considered as a multi-degree of freedom (MDOF) vibration system (see Sect. 1.4).

1.1   Kinematic of Vibrations

We discuss the kinematic of vibrations independent from the fact which sources the vibrations have and at which locations of a system they occur. We will concentrate on periodic vibrations and the special case of harmonic vibrations, because of their dominance in vibration problems of power plant machines.

1.1.1   Periodic and Harmonic Vibration Signals in the Time Domain

Figures 1.1 and 1.2 describe three typical periodic oscillations $s(t)$: triangular, sinusoidal and rectangular. An important example of them is the harmonic sinusoidal time function:

$$s(t) = s_{max} \sin \omega t \tag{1.1}$$

with the amplitude s_{max} and the angular frequency ω.

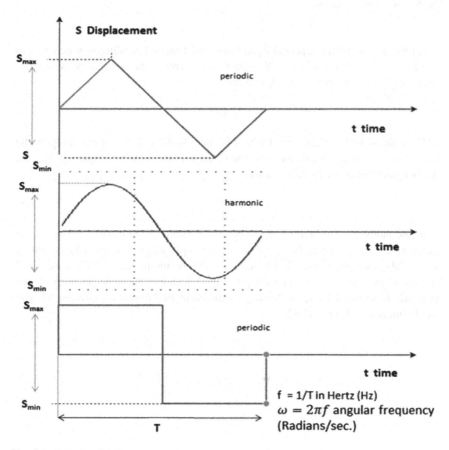

Fig. 1.1 Periodic vibrations

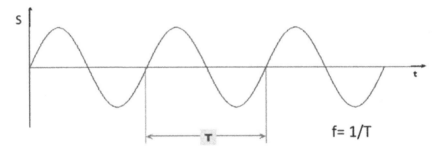

S: displacement   T: time of one period
f: frequency in Hertz   t: time

Fig. 1.2 Harmonic vibration

After elapse of the time period T, all three time function conditions which govern the changes repeat themselves. We can now determine the oscillatory (vibration) frequency f from the period T, as follows:

$$f = 1/T \tag{1.2}$$

The unit of the frequency f is Hertz: f is equal to the number of events (periods) per second. In vibration calculations, one often uses the angular frequency, i.e., the number of oscillations in 2π seconds:

$$\omega = 2\pi f \tag{1.3}$$

In the analysis of vibration, the subject of "harmonic motion" is of particular importance. This is when the variables which apply change in accordance with a sinusoidal curve (see Figs. 1.1, 1.2 and 1.3). The harmonic motion is characterized by one frequency only. As an example, Fig. 1.3 shows the superposition of two harmonic functions S1 and S2 leading to the combined periodic vibration S3 with two frequencies of S1 and S2.

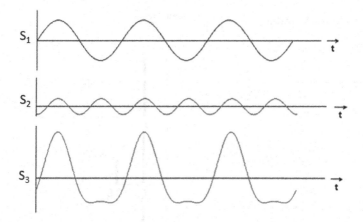

Fig. 1.3 Combined vibrations

1.1.2 Vibrations in the Time and Frequency Domain (Fourier Analysis)

The Fourier analysis is applied, if the different frequencies being involved in a vibration event. *The vibration event should be individually recognized.* We speak about frequency analyzers or Fourier analyzers. These instruments have a special importance at modal analysis applications like in structural resonance problems. They enable a transition from the time domain into the frequency domain.

Every periodic oscillation can be traced back to a combination of harmonic (i.e., sinusoidal) oscillations. Considering as example a rectangular form (see Fig. 1.4) and using the Fourier transform method, this time function $s = f(t)$ can be transformed from the time domain to the frequency domain $s = f(f)$ and we obtain a spectrum. Figure 1.4 indicates the Fourier transformation as a transition from time to frequency domain:

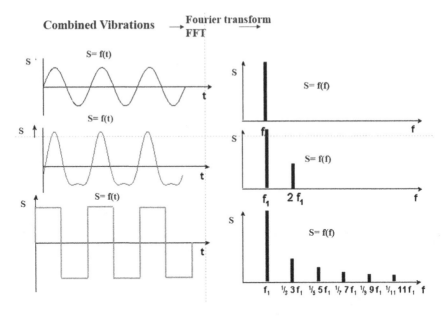

Fig. 1.4 Time-frequency functions

If the rectangular oscillation is given by the time period T and the basic frequency $f = f_0 = 1/T$ in the time domain, one finds in the frequency domain that this is a combination of basic frequency f_0 and an infinite number of harmonic components $f_1, f_2, f_3, \ldots, f_\infty$.

Harmonic frequency components f_1 to f_∞ are odd-numbered multiples of the fundamental f_0:

$$f_1 = 1f_0, f_2 = 0, f_3 = 3f_0, f_4 = 0, f_5 = 5f_0, \ldots$$

If we look at a triangular oscillation, we will find even-numbered frequency components:

$$f_1 = f_0, f_2 = 2f_0, f_3 = 0, f_4 = 4f_0, f_5 = 0, f_6 = 6f_0, \ldots$$

The Fourier cube is a visualization of the time domain and the frequency domain in a 3-dimensional presentation (see Fig. 1.5):

- From the frequency view, the different frequency lines are visible as a spectrum.
- From the time view, the superimposed time functions of these frequencies are visible.

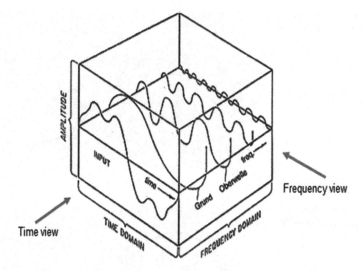

Fig. 1.5 Fourier cube

The following Figs. 1.6, 1.7, 1.8, 1.9 and 1.10 demonstrate how a periodic function is built from harmonics.

Fig. 1.6 Fundamental (Brüel and Kjäer Vibro 1995)

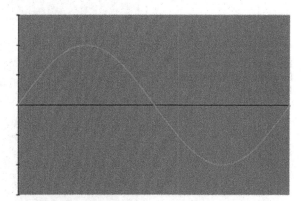

Fig. 1.7 Third harmonic added (Brüel and Kjäer Vibro 1995)

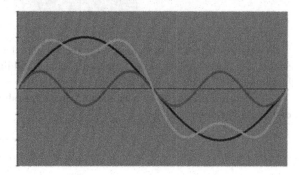

Fig. 1.8 Fifth harmonic added (Brüel and Kjäer Vibro 1995)

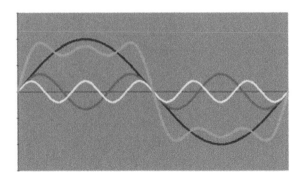

Fig. 1.9 Seventh harmonic added (Brüel and Kjäer Vibro 1995)

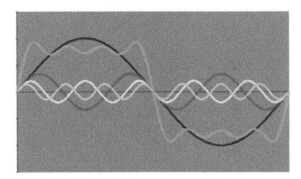

Fig. 1.10 Fifth to ninth harmonic added (Brüel and Kjäer Vibro 1995)

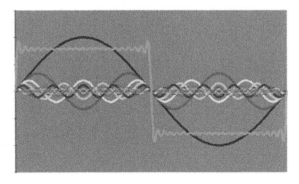

We are now already very close to the rectangular shape. As more harmonics are added, the closer we get to the rectangular shape. Adding an indefinite number of harmonics will result in an ideal rectangular signal.

The Fourier rule says: All periodic and quasi-periodic signals are a combination of several harmonic signals.

1.1.3 Relations Between Deflections, Velocities and Accelerations

We consider a simple mass-spring vibration system as shown in Fig. 1.11 and assume that the point-mass m performs a harmonic up and down movement, where the deflection $s(t)$ follows a time function as described in Eq. (1.4). It describes a motion whereby its distance from the zero position varies according to a sinusoidal time function. Such a harmonic motion of the SDOF system can occur either as a free motion after a short disturbance or as a forced motion due to some harmonic excitation. Solutions can be found from the equations of motion for the SDOF system as it will be derived in the next Sect. 1.2. From a kinematic point of view, the harmonic time function shown in Fig. 1.2 is determined by the projection of an arrow, rotating in a polar diagram with the angular of velocity ω (see Fig. 1.11 for the definition quantities of a spring pendulum).

The origin of the polar graph is 0, but we define an arbitrary chosen starting point $t = 0$ for our considerations. This because of the balancing phase reference, which will be explained later.

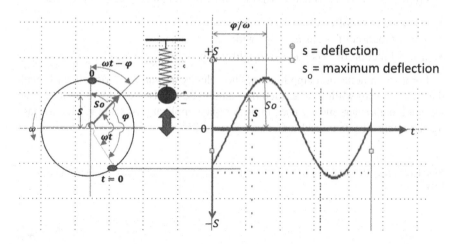

s = deflection
s_o = maximum deflection

Fig. 1.11 Mass on a spring

If we consider the period starting at $t = 0$, the relationship between s and t is given by the following equation:

$$s = s_0 \cos(\omega t - \varphi) \tag{1.4}$$

where s_0 is the deflection or vibration amplitude and φ is the phase angle. Both quantities are very important, particularly when different time signals must be superimposed.

We will need the phase angle later for balancing, which will be described in more detail in Chap. 4. The phase angle φ is determined from the point of time zero. With most of the vibration measurements made on turbo-sets, time zero is set by a reference signal, generated by a photo-electric or magnetic pickup from a mark on the shaft, so that one impulse peak is given for each revolution.

The balancing phase reference will be explained later.

So far, we have been considering the deflection s which is also referred to as the vibration amplitude. If the deflection s is now differentiated with respect to time, we obtain the velocity v:

$$v = \frac{\mathrm{d}s}{\mathrm{d}t} = -s_0\omega \sin(\omega t - \varphi) = v_0 \sin(\omega t - \varphi) \qquad (1.5)$$

If Eq. (1.4) is now differentiated a second time with respect to time, we obtain the vibration acceleration a:

$$a = \frac{\mathrm{d}v}{\mathrm{d}t} = \frac{\mathrm{d}^2 s}{\mathrm{d}t^2} = -s_0\omega^2 \cos(\omega t - \varphi) = a_0 \cos(\omega t - \varphi) \qquad (1.6)$$

From Eqs. (1.4), (1.5) and (1.6), the following relationship becomes clear:

- s is a function of $\cos(\omega t - \varphi)$ with factor s_0
- v is a function of $-\sin(\omega t - \varphi)$ with factor $s_0\omega$
- a is a function of $-\cos(\omega t - \varphi)$ with factor $s_0\omega^2$.

For a given deflection amplitude s_0, the velocity amplitude v_0 rises linearly with the frequency ω while the acceleration amplitude a_0 rises quadratic with ω. The vibration parameters mostly used in power plant applications, s and v, have a phase displacement of 90° to each other (see Fig. 1.12).

Fig. 1.12 Phase relation between s and v

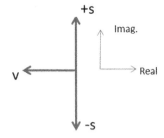

In practice, oscillation or vibration takes place when a mass is subject to forces under resilient conditions. Mass and spring elements are the requisite components of a system which is capable of vibration or oscillation. The vibration may either be free (i.e., natural) or forced vibration.

During the motion, spring forces, mass inertia forces and external forces are acting, and at each instant are in a state of equilibrium. In actual practice, another element is also present in the form of frictional forces, which act in opposition to the direction of motion and which must be included in the conditions of equilibrium. The effect of friction is called damping. If its effect is so small that it can be neglected for the considered case, we have an undamped (or weakly damped) system, otherwise it is damped (see Fig. 1.13).

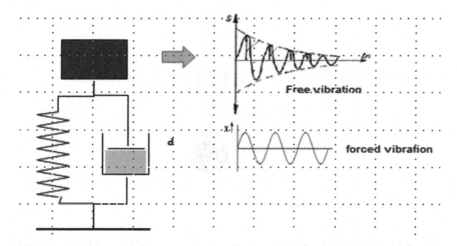

Fig. 1.13 Damped vibration

In contrary to the free, undamped vibration, the damping causes that the system will restore the standstill according to an exponential function. How long that takes depends on the amount of damping. A mass-spring system comes into free vibration if the mass is once displaced from its resting position and then allowed to move on its own; forced vibration takes place if the system is continuously kept in motion by an external force.

In Fig. 1.13, we see the difference between a free and a forced vibration. In practice, a free vibration will decay because of the damping d. A forced vibration will be kept up, because of an external driving force.

1.2 Vibrations of a Single Degree of Freedom (SDOF) System

A very simple mechanical system to explain the basics of vibrations is the single degree of freedom (SDOF) system, consisting of the two parameters mass m and spring c. For this basic system, the equations of motion, the natural frequency, the free vibrations and the forced vibrations will be derived and discussed in this chapter. The additional effect of damping on free vibrations and forced vibrations will also be investigated.

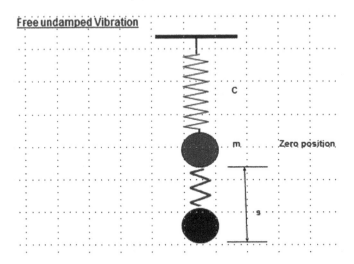

Fig. 1.14 Free vibration of a SDOF system

Free undamped vibrations: A ball of mass m as shown in Fig. 1.14 is suspended from a linear spring having the spring constant c. The ball is next displaced by amount s and then allowed to move freely. We now want to find the natural frequency of the oscillation of the mass-spring system, in other words the frequency of the system, at which "resonance" appears. In case of the free vibration, the forces acting on the ball are:

- the spring restoring force $F = -cs$ and
- the inertia force of the mass m.

The minus sign indicates that F acts in the opposite direction to the deflection s. In accordance with the fundamental law of dynamics, expressed by Newton's law, these two forces must be in equilibrium:

$$-cs = m\frac{d^2s}{dt^2} \Rightarrow \frac{d^2s}{dt^2} + \frac{c}{m}s = 0 \tag{1.7}$$

For this equation of motion for the free vibrations, the time solution can be obtained by $s = s_0 \sin \omega t$ and with the derived acceleration:

$$\frac{d^2s}{dt^2} = -\omega^2 s = a \tag{1.8}$$

we obtain the angular natural frequency ω of the SDOF system:

$$\omega^2 s = \frac{c}{m} s \Rightarrow \omega = \omega_0 = \sqrt{\frac{c}{m}} \tag{1.9}$$

The natural frequency can also be expressed in Hertz. This natural frequency is a function of c and m.

$$f = \frac{1}{2\pi} \sqrt{\frac{c}{m}} \tag{1.10}$$

Forced undamped vibration: If the mass-spring system is caused to vibrate by external means, the ball will be subject to forced vibration. We consider the special case of a forced ground excitation as shown in Fig. 1.15.

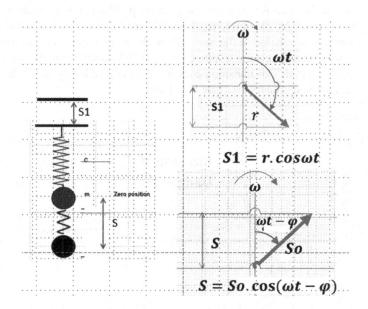

Fig. 1.15 Forced vibration due to ground excitation

We concentrate on the "vibrating condition" after all initiating processes have died out. If the point of suspension (or of excitation) has moved by amount s_1, the spring has been lengthened by amount $s - s_1$ and as per the basic law of dynamics, we find the equations of motion for the forced vibration of the SDOF:

$$m\frac{d^2s}{dt^2} + cs = cs_1 \tag{1.11}$$

Substituting s_1, we obtain the equation of motion:

$$m\frac{d^2s}{dt^2} + cs = cr\cos(\omega t) \tag{1.12}$$

By solving Eq. (1.12) for the deflection s, we obtain:

$$s = \frac{cr}{c - m\omega^2}\cos(\omega t) \tag{1.13}$$

Thus, the deflection depends on the frequency ω. At very low values of ω, the maximum value becomes practically equal to r, and the forced vibration hardly differs from the exciting oscillation. We could consider that as the "rigid state." When ω reaches the value $\omega = \omega_0$ (the case of resonance), the denominator becomes zero and s approaches infinity. In practice, infinite deflection does not occur, because it is damped by damping effects which are always present. When ω is increased further, above, the denominator becomes negative; this means that the excitation point reverses direction in relation to the mass. Thus, the mass experiences, as it passes through the resonance frequency, a "phase jump" of 180° in relation to the movement of the suspension point. This transition does not take place suddenly, as the damping results in a continuous changeover (see Fig. 1.16).

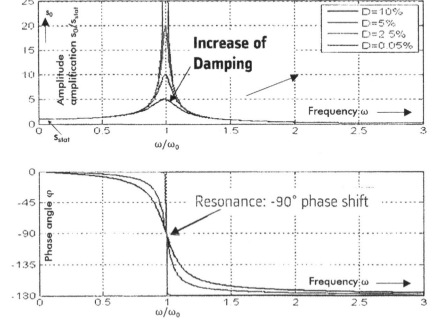

Fig. 1.16 Resonance, phase shift, damping

Figure 1.16 demonstrates the effect of damping, expressed by the damping factor *D*. Damping sources can be

- material,
- joints,
- journal bearings,
- seals and
- electromagnetic interaction.

Below its resonance frequency, a flexible system is governed by its spring properties $1/c$. At the high point of the resonance, the damping will limit the response $1/\omega d$. After having passed the resonance, the inertia mass will govern the response.

The effect of damping on the free as well the forced vibrations will be derived in the following (Sects. 1.2.1 and 1.2.2).

1.2.1 Effect of Damping on Free Vibration

Up to now, the damping effect was neglected in the equations of motion, as shown again for the free vibrations

$$m\frac{d^2s}{dt^2} + cs = 0 \tag{1.14}$$

where $m\frac{d^2s}{dt^2}$ represents the mass inertia force and cs represents the spring restoring force.

As described earlier, in real cases there are always forces present which have a damping effect. The damping forces perform work and reduce the content of kinetic energy in the system. To illustrate this, we will add a damping cylinder to the simple model, where the piston follows the same motion as the mass (see Fig. 1.17).

Fig. 1.17 Damped spring-mass system

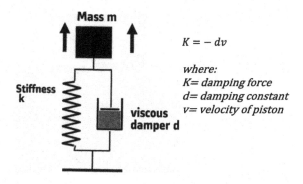

Mass m

Stiffness k

viscous damper d

$K = -dv$

where:
K = damping force
d = damping constant
v = velocity of piston

The sign of the damping force is negative because the damping force acts against the motion and slows it down. Again, the dynamic forces are in equilibrium for the free vibrations, so that:

$$m\frac{\mathrm{d}^2 s}{\mathrm{d}t^2} + d\frac{\mathrm{d}s}{\mathrm{d}t} + cs = 0 \tag{1.15}$$

In the cylinder, frictional forces K occur which act in the opposite direction to the vibration movement and which have a magnitude proportional to the velocity v of the piston (in this case one speaks of a "linear damping," which is the usual situation in our work).

Equation (1.15) is the equation of motion for the damped, free vibration. The terms $m\frac{\mathrm{d}^2 s}{\mathrm{d}t^2}$ and cs are known from the previous discussion, and the term $d\frac{\mathrm{d}s}{\mathrm{d}t}$ now represents the damping force.

When solving this equation for the displacement s by means of a mathematical set up, we obtain

$$s = s_0 e^{-DT}\cos(\omega t - \varphi) \tag{1.16}$$

where D is the constant for decay time, in the technical literature known as damping measure:

$$D = \frac{d}{2\sqrt{cm}} \tag{1.17}$$

D is a dimensionless number, which in our practice lies between 0 and 1 (or between 0 and 100%). A good way to determine D is the "logarithmic decrement δ," by consideration of the process of decay.

The initially undamped time function $s = s_0\cos(\omega t - \varphi)$ will be enveloped by the damping $s = s_0 e^{-DT}$ and decays logarithmically. When the values of the vibration maxima are plotted on a vertically scaled logarithmic graph, each maximum of the points plotted must lie on a straight line (see Fig. 1.18).

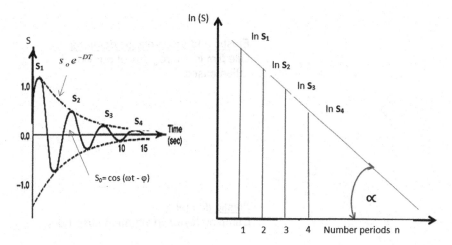

Fig. 1.18 Decay rate by damping

The tangent of angle α, which is the angle of slope of the line, is the desired logarithmic decrement δ:

$$\delta = \tan \alpha = \frac{\ln s_1 + \ln(s_n + 1)}{n} \tag{1.18}$$

The damping measure D can now be determined from δ as follows:

$$D = \frac{\delta}{\sqrt{4\pi^2 + \delta^2}} \tag{1.19}$$

The damping in our machinery applications is the linear velocity-proportional (viscoelastic) damping. The main active damping at overcritical rotors at turbo-sets is provided by the bearings and is around $D = 3\text{--}5\%$. At overcritical rotors, the internal material damping force will be antiphase to the external bearing (oil film) damping and is therefore destabilizing.

The following Fig. 1.19 shows different kinds of damping characteristics.

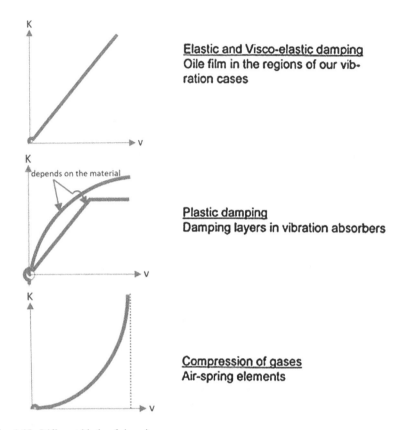

Fig. 1.19 Different kinds of damping

1.2.2 Effect of Damping on Forced Vibration

Just as with free vibration, the forces in the system now including also the damping force must remain in equilibrium. Hence, the equation of motion is now

$$m\frac{\mathrm{d}^2 s}{\mathrm{d}t^2} + d\frac{\mathrm{d}s}{\mathrm{d}t} + cs = cr\cos(\omega t) \qquad (1.20)$$

The terms on the left side of Eq. (1.20) are already known as inertia force, damping force and spring restoring force; the new element is term $cr\cos(\omega t)$ which represents the excitation force at the point of excitation or driving point of the system.

The oscillating mass may well have the same frequency as the excitation point, but not the same phase. There is a phase difference φ, and the point of time zero is determined by the excitation function. The forced vibration lags the excitation, and the phase angle φ depends on the angular frequency and on damping constant d.

From the equation of motion (see Eq. 1.20), the maximum displacement s_0 can be derived:

$$s_0 = \frac{cr}{\sqrt{(c - m\omega^2)^2 + d^2\omega^2}}$$ (1.21)

and for the phase angle φ:

$$\tan \varphi = \frac{d\omega}{c - m\omega^2}$$ (1.22)

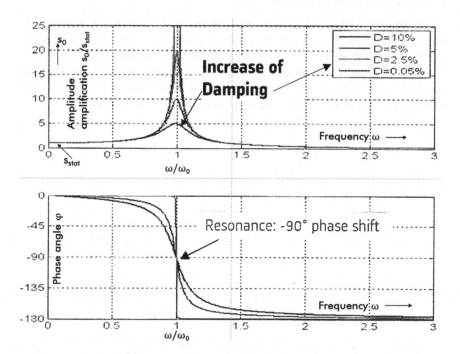

Fig. 1.20 Change of amplitude by frequency and damping

The deflection of the oscillating or vibrating body thus varies with the angular frequency and the damping. Figure 1.20 shows the relationship between the two values φ and the damping in the range of $D = 0.05$–10%.

1.3 Vibrations of a Simple Rotating Shaft—The Laval Rotor

The Laval rotor is named according to Gustav de Laval, a Swedish inventor. He invented the Laval turbine in 1883: a fast rotating flexible shaft and the shaft resonances (critical speeds) below the operating speed.

Let us now consider a simple rotating shaft system. We assume that

- the shaft is flexible and has external damping,
- its mass is concentrated in a disc at its midpoint so that the rest of the shaft has no mass but simply acts as a massless spring and
- unbalance is assumed in the disc only.

This theoretical Laval rotor can move in the two directions. It is therefore, strongly considered, a two degree of freedom (TDOF) system: one in the y direction and the other one in the z direction, the direction of weight G (see Fig. 1.21). However, if there is no coupling between the two directions, we can treat the vibration problem as two SDOF systems. Each of this SDOF system has one resonance, which is also called critical speed (Figs. 1.22 and 1.23).

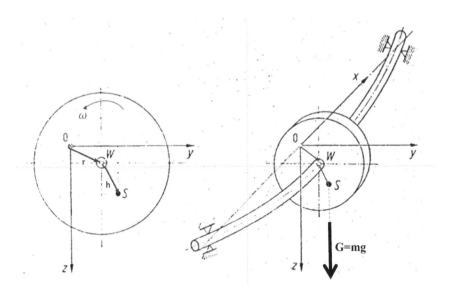

Fig. 1.21 Laval rotor (Gasch and Pfützner 1975)

The given data are the rotational frequency ω, the gravitational acceleration g, the disc mass m and shaft spring constant c, and the deflection h of center of gravity S from shaft midpoint W, which is deflected by the unbalance.

What we wish to determine is the deflection $r(t)$ of the midpoint W of the shaft. At any given instant, the forces acting must be in equilibrium and the torques must be in balance. The calculation of the balance of moments, however, is unnecessary, because:

- we are considering steady-state operation (ω is constant) and
- the shaft deflection r or eccentricity h are very small in relation to the radius of inertia of the rotating shaft (radius of inertia $= \sqrt{\text{mass moment of inertia mass}}$).

The occurring displacements are defined as follows:

i. displacement of center of disc W: r_y and r_z
ii. displacement of center of gravity S: h_y and h_z.

The two equations of motion for the two degrees of freedom are:

$$y\,\text{direction:}\quad m\ddot{r}_y + cr_y = mh\omega^2 \cos(\omega t) \tag{1.23}$$

$$z\,\text{direction:}\quad m\ddot{r}_z + cr_z = mh\omega^2 \sin(\omega t) \tag{1.24}$$

The solutions for the two degrees of freedom are:

$$r_y(t) = A_y \cos \omega_0 t + B_y \sin \omega_0 t + h\frac{\left(\frac{\omega}{\omega_0}\right)^2}{1 - \left(\frac{\omega}{\omega_0}\right)^2}\cos(\omega t) \tag{1.25}$$

$$r_z(t) = A_z \cos \omega_0 t + B_z \sin \omega_0 t + h\frac{\left(\frac{\omega}{\omega_0}\right)^2}{1 - \left(\frac{\omega}{\omega_0}\right)^2}\sin(\omega t) + \frac{G}{c}$$

Homogeneous solution:	Inhomogeneous solution:		(1.26)
Natural vibrations with natural frequency: $$\omega_0 = \sqrt{\frac{c}{m}}$$ This part usually decays due to damping.	Forced unbalance response with circular frequency ω which is the angular velocity. Vibration amplitude depends on h and $\frac{\omega}{\omega_0}$.	Static deflection due to weight G.	

Influence of external damping: The damping of the shaft is only provided by the bearings. For the rotating shaft, the following is valid:

$$\text{mass moment of inertia} = \text{mass}*(\text{radius of inertia})^2$$

Fig. 1.22 Deflection and
damping of the Laval shaft

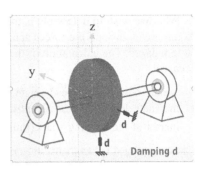

The two equations of motion, including the influence of external damping, for
the two degrees of freedom are:

$$m\ddot{r}_y + d\dot{r}_y + cr_y = mh\omega^2 \cos(\omega t) \tag{1.27}$$

$$m\ddot{r}_z + d\dot{r}_z + cr_z = mh\omega^2 \sin(\omega t) \tag{1.28}$$

Natural frequency of the damped system:

$$\omega_d = \omega_0\sqrt{1 - D^2} \tag{1.29}$$

$$\omega_0 = \sqrt{c/m} \tag{1.30}$$

$$D = d/2\sqrt{cm} \tag{1.31}$$

Fig. 1.23 Influence of
different external dampings

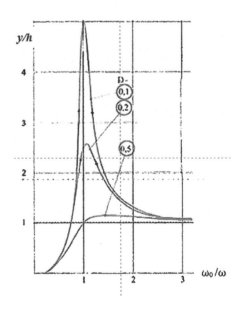

1.4 Rotors in Practice

Adding another mass disc to the shaft will create an additional critical speed. The same happens if you would add a third mass disc to the shaft. The practical rotors have a continuous distribution of masses and springs along its axial length. The consequence will be an indefinite number of critical speeds.

Fig. 1.24 Mode shapes at critical speeds

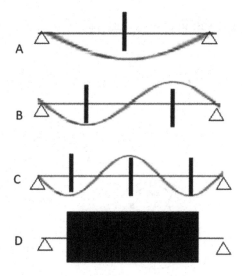

A: 1 mass + 1 mass-less spring (shaft) = 1 critical speed
B: 2 masses + 1 mass-less spring (shaft) = 2 critical speeds
C: 3 masses + 1 mass-less spring (shaft) = 3 critical speeds
D: Continuous distribution of masses and spring properties along the rotor axis results in an indefinite number of critical speeds.

When the rotor passes its 1st resonance (critical speed A in Fig. 1.24), a transition from the rigid into the elastic state takes place, manifested by a lagging phase change of 180° of the responding vibration r to the exciting unbalance h which is the eccentricity to the center of gravidity S.

At the resonance high point, the (maximum) vibration will lag 90° behind the unbalance force. Having passed the resonance, the shaft is self-centering.

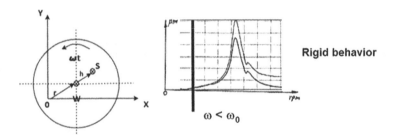

Fig. 1.25 Rigid status of the rotor

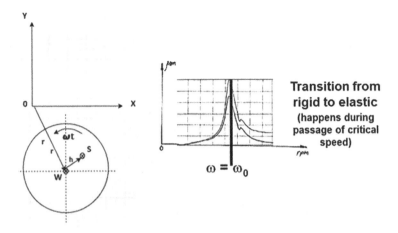

Fig. 1.26 High point of critical speed, transition to flexible

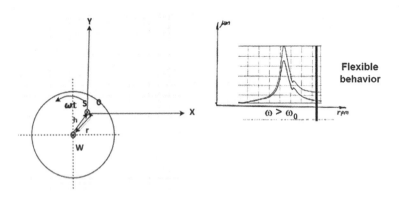

Fig. 1.27 Flexible status of the rotor

Figures 1.25, 1.26 and 1.27 demonstrate the transition of the rotor from rigid to flexible:

- In Fig. 1.25, the operating speed ω is below the critical speed ω_0 and no phase difference between the unbalance force h and the resulting deflection.
- Passing the 1st critical, the transition to flexible takes place and at the high point (where the operating speed ω corresponds to the critical speed ω_0), the phase angle between h and r will be 90° (see Fig. 1.26).
- At higher operating speed than the critical speed, there is a low deflection again, and h and r are in antiphase (see Fig. 1.27).

Gas turbine rotors, HP and IP steam turbine rotors normally pass the 1st and sometimes the 2nd critical speed, before they reach operating speed, and large generator rotors will have up to three critical speeds below the operating speed. As shown in Fig. 1.24, the vibration phase is kept throughout A, B and C on one end of the rotor, but it changes on the other end. It will be kept on the end where the excitation force is induced, and it changes on the "free" end.

Fig. 1.28 Turbomachinery bending modes occurring in practice (Brüel and Kjäer Vibro 2002)

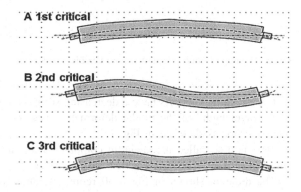

A 1st critical

B 2nd critical

C 3rd critical

At most of turbomachinery rotors, the bending modes according to Fig. 1.28 will occur within the operating speed range.

If the bearings are infinitely stiff, the nodal points of the modal deformations are forced into the supporting points. If the supporting points (bearings) have a certain limited stiffness, they will vibrate, and the nodal points are travelling toward outside. This causes the critical speed to drop.

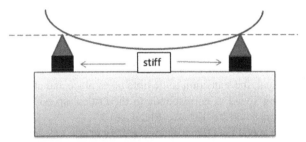

Fig. 1.28. critical speed stiff

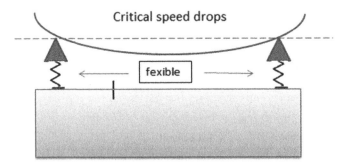

Fig. 1.29 Drop of critical speed by more flexible bearings

If the bearing supports have a decreasing stiffness, the critical speeds will drop: The "wavelength" of the modal deformation becomes bigger; therefore, the frequency drops as sketched in Fig. 1.29.

Next is a so-called n/α diagram combined with calculated mode shapes (n: = speed in rpm, α: = bearing elasticity in mm/MN).

The diagram shown in Fig. 1.30 represents an example of the critical speed calculation of a large steam turbine in Poland. You can see the dependency of the critical speed upon the bearing stiffness in mm/MN on the left diagram and the mode shapes on the right.

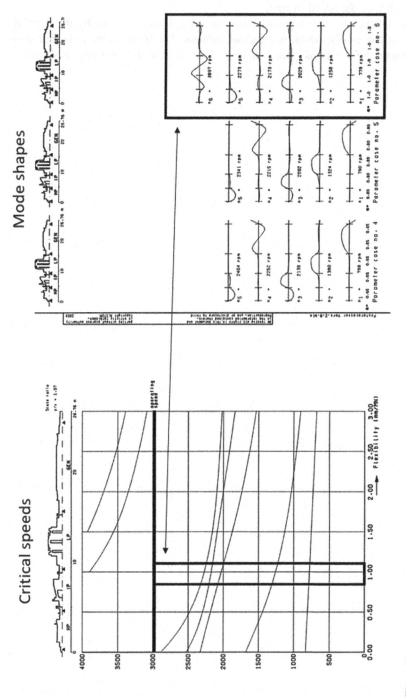

Fig. 1.30 n/α diagram of a 150 MW steam turbine

The horizontal axis represents the bearing stiffness in mm/MN, and the vertical axis indicates the speed in rpm.

The left part of the diagram shows the critical speeds, whereby the there is a frame at 1 mm/MN, which is a normal stiffness. The frame in the right part shows the kinetic deformations (mode shapes) of the shaft at different critical speeds.

Chapter 2
Instrumentation and Measurement

For vibration analysis and troubleshooting, nowadays almost worldwide frequency analyzers based on Fast Fourier Transformers (FFT) are used. They will work in the time mode, producing time functions without FFT, or in the frequency mode, producing frequency functions with FFT (see also Sect. 1.1.2). All kinds of vibration transducers can be connected in order to visualize and analyze the vibration events under investigation.

This chapter describes the most usual analyzing instruments, their visualization techniques, transducer types and the usual measuring positions.

2.1 Measuring and Analyzing Instruments

In the last decades of our career, the vibration instruments of Bently Nevada have almost become a worldwide standard. Figure 2.1 shows the elder ADRE FOR WINDOWS measuring system (Type 208) by Bently Nevada, which is used by vibration engineers worldwide. It consists of one or two 8-channel front end units (DAIU), so it covers max. 16 channels. It can handle two independent shafts, having two key-phasor inputs. The MMC unit is normally a laptop.

© Springer Nature Switzerland AG 2020
F. Herz and R. Nordmann, *Vibrations of Power Plant Machines*,
https://doi.org/10.1007/978-3-030-37344-3_2

Fig. 2.1 Old Adre 8 system

Nowadays, mostly the later version, the ADRE 408 as displayed in Fig. 2.2, will be used together with the ADRE SXP software package. This software can handle the data from the 208 as well as the 408 system. One 408 chassis handles up to 32 channels, and it can handle 3 to 6 independent shafts having six key-phasor inputs.

Fig. 2.2 408 Adre system (Protective Supplies & Procurement Services 2019)

In the power plant applications, we normally use:

- displacement for shaft vibrations,
- velocity for bearing cap or structural (foundation) vibrations and
- acceleration for high-frequency applications like roller bearing or gearbox applications.

Measurement types for mechanical vibrations:

- Vibration displacement s in μm or mil
- Vibration velocity v in mm/s or inch/s
- Vibration acceleration a in m/s^2 or g (1 mil = 25.4 μm, 1 in/s = 25.4 mm/s).

The relative shaft vibration between rotating shaft and pedestal are measured with proximity probes (eddy current probes). The absolute bearing vibrations are measured on the pedestal with seismic or velocity probes. Occasionally, the absolute shaft vibrations are also measured: they are either determined from the relative shaft vibration and the pedestal vibrations or they are measured by means of shaft gliders and velocity probes at elder installations.

In Fig. 2.3, the usual measuring types in most turbine power plants are shown.

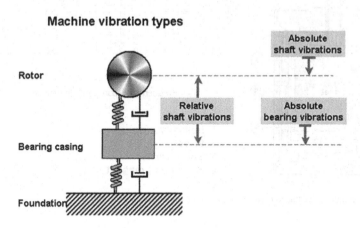

Fig. 2.3 Usual vibration measuring types in turbine applications of power plants (Brüel and Kjäer Vibro 1995)

2.1.1 Velocity Transducers

The spool, the spool carrier (mass) and the helical spring form together a highly damped, seismic suspended spring-mass system.

The resonance frequency of the system is around 10 Hz. If the transducer is exposed to vibrations, there will be relative motion of the spool against the magnet above 10 Hz. In case of relative motion, a voltage will be induced in the spool, which will be proportional to the velocity according to the law of induction, since the voltage equals to the magnetic field strength times spool length times cutting velocity of the magnetic field.

Below 10 Hz, we will have reduced or no relative motion and no output from the spool. To enable a working range lower than the seismic resonance of 10 Hz, there is an electronic linearization which flattens and phase-corrects the response (see Fig. 2.4).

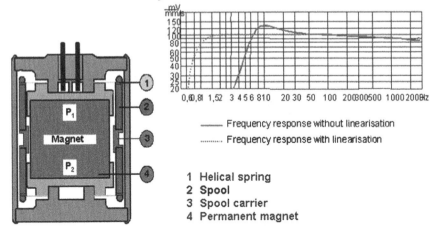

**Absolute bearing vibrations -
Vibration velocity sensor**

1 Helical spring
2 Spool
3 Spool carrier
4 Permanent magnet

Fig. 2.4 Function of the velocity sensor (Brüel and Kjäer Vibro 1995)

The advantages are:

- rugged design,
- high sensitivity even at low frequencies,
- high output signal with low internal resistance,
- requires no external power and
- oil-, water-, vacuum-tight and chemically resistant (stainless-steel casing).

The drawbacks are:

- Upper frequency limited to 2000 Hz
- Influenced by strong magnetic fields
- Resonance at low frequency measurement (linearization is needed)
- Relatively large size.

Figure 2.5 shows the transducer types used in the power plants (Brüel and Kjäer Vibro 1995).

Fig. 2.5 The mostly used
velocity transducers (Brüel
and Kjäer Vibro 1995)

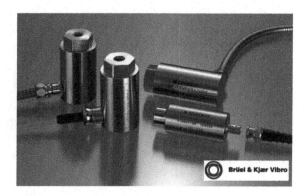

2.1.2   Eddy Current Measuring Chain (Proximity Transducers)

An eddy current measuring chain (or proximity transducer) consists of an oscillator and a transducer. In between there is a specially tuned cable to ensure the correct sensitivity value.

The oscillator produces an alternating signal (~ 1 MHz) which creates an alternating magnetic field. In electrically conducting elements close (in proximity) to the sensor, eddy currents are induced. They set up a magnetic field in a direction opposing the magnetic field that creates them. As closer the distance to the transducer is, as lower the voltage output of the transducer will be. So, the eddy current probe measures distances and variation of distances (vibrations).

There must be a certain distance or gap between the transducer and the specimen to be measured (normally ~ 2 mm). This distance results in a proportional output voltage (gap voltage: -8 to -11 V) for instance at a standing shaft. As soon as the shaft begins to rotate, the shaft vibrations appear as an overlapping over the gap voltage. This transducer signal is very useful for measuring shaft vibrations, because it contains the information about run-out as well as vibration.

Figure 2.6 shows the commonly proximity transducer used for permanent installation in power plants. The transducer in Fig. 2.7 will be used for temporary installations, such as trouble shooting.

Fig. 2.6 Eddy current
transducer and oscillator
(Brüel and Kjäer Vibro 1995)

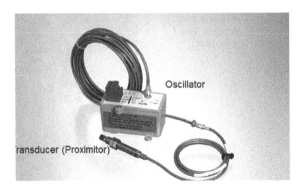

Fig. 2.7 Eddy current sensor
with integrated oscillator
(Brüel and Kjäer Vibro 1995)

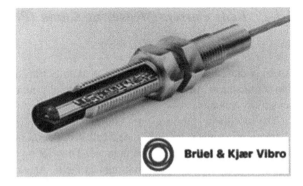

2.1.3 Acceleration Sensors (Accelerometer)

The sensing elements in acceleration sensors (accelerometers) are piezo-electric
discs (element no. 5 in Fig. 2.8), which create an electric charge as soon as they are
exposed to pressure. The seismic mass discs (element no. 4 in Fig. 2.8), attached
with the pretensioning element discs (element no. 3 in Fig. 2.8), create an alternating pressure because of its inertia. The pressure (or force) equals to the mass
times acceleration. The mass (no. 4 in Fig. 2.8) is given and constant so the output
signal is proportional to acceleration. The pressure variations cause charge variations, which are transformed into voltage variations by means of the built-in charge
amplifier. The resonance frequency is very high (~ 30 kHz) and the frequency
response is almost flat.

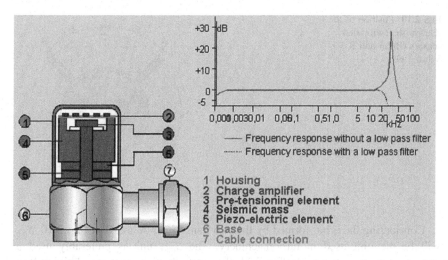

Fig. 2.8 Principle of an accelerometer (Brüel and Kjäer Vibro 1995)

The impedance of an accelerometer is very high (in the MΩ range); this is one of the reasons that it is rarely used at the main units. It is used at higher frequency ranges, e.g., for roller bearing measurements at auxiliary machines and pumps.

Figure 2.9 shows different designs of accelerometers and a charge amplifier. The charge amplifier is necessary to convert the electric charge into a voltage.

Fig. 2.9 Accelerometers (Brüel and Kjäer Vibro 1995)

2.2 Measuring Positions and Units

The shaft vibration is a relative displacement vibration between bearing pedestal and shaft. The measurement of this relative displacement is done with contactless eddy current sensors (see Sect. 2.1.2). The shaft vibration eddy current probes of the main units are usually positioned as shown in Fig. 2.10.

Fig. 2.10 Position of the
relative shaft vibration
sensors (Brüel and Kjäer
Vibro 1995)

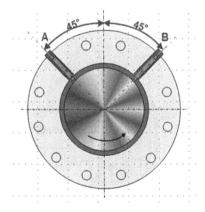

Considering the orbit, formed by the two transducers A and B in Fig. 2.10, $S_{p.p.}$ will be the unit used in turbomachinery. The S_{max} unit is preferably used at low-speed hydromachines. In Fig. 2.11, the shaft vibration quantities deriving from the orbit can be seen.

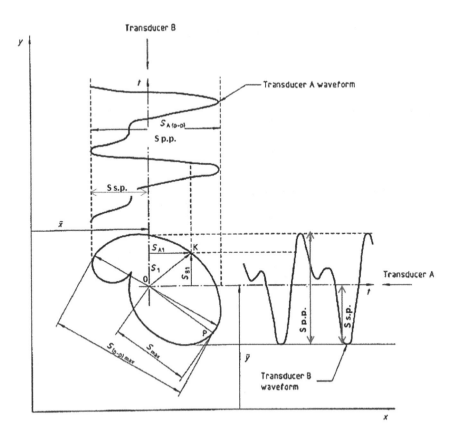

Fig. 2.11 Forming the orbit of sensor A and B (International Organization for Standardization 1996)

The pedestal vibration is measured as an absolute vibration velocity in mm/s_{rms}. Here, velocity transducers are used (see Sect. 2.1.1). Usually, only the vertical vibrations are measured, and some plants install the horizontal measurement as well (see Fig. 2.12).

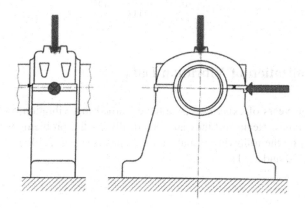

Fig. 2.12 Measurement locations for pedestal vibrations (International Organization for Standardization 1996)

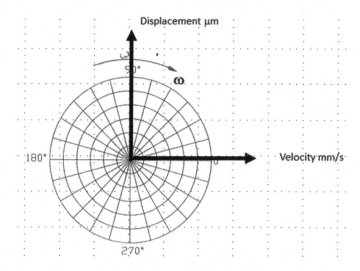

Fig. 2.13 Phase difference between displacement and velocity

To obtain velocity from displacement, we must differentiate one time. That means that a $\cos(\omega t)$ becomes a $\sin(\omega t)$ with is a phase shift of 90° (Fig. 2.13).

Here is a quick formula to transfer velocity to displacement:

$$s_{\text{p.p.}}(\mu m) = \frac{v\left(\frac{mm}{s}\right)}{f(\text{Hz})} \cdot 450 \tag{2.1}$$

2.3 Visualization of Vibration Data

During my 50 years of experience, we have learned that vibration problems, covering gas turbines, steam turbines and practically all the problems we had faced, were covered by the following visualization graphs (see Fig. 2.14, 2.15, 2.16, 2.17, 2.18, 2.19, 2.20 and 2.21).

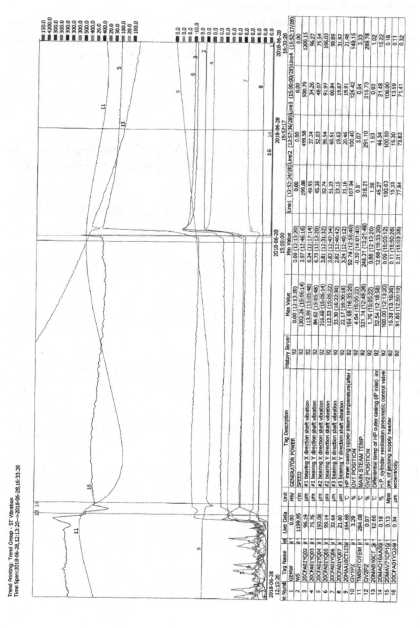

Fig. 2.14 Surveillance plot (time trend) of permanently installed surveillance

The vertical axis of the graph systems shows the chosen machine parameters with the scaling parameters on its right side. The table below shows the chosen parameters. Practically all the important parameters (temperatures, pressures, flow values, etc.) can be chosen for display.

In Fig. 2.14, the vibrations had been chosen, besides the operational main parameters. The visualization of vibrations is usually used for trouble shooting purposes (these graphs had been taken from the Adre SXP program).

2.3.1 Time Trend

Figure 2.15 shows the time trend figure of the Adre SXP analyzing system. The upper graph represents the integrated phase angle regarding the reference signal.

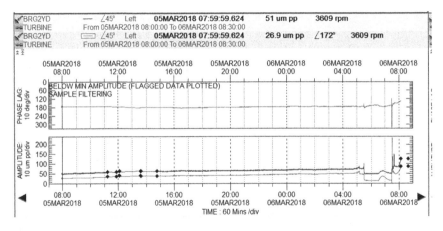

Fig. 2.15 Time trend from SXP

2.3.2 The Bode Plot

The Bode plot, which is a double graph, indicates the amplitude and the phase angle in dependency of the machine speed.

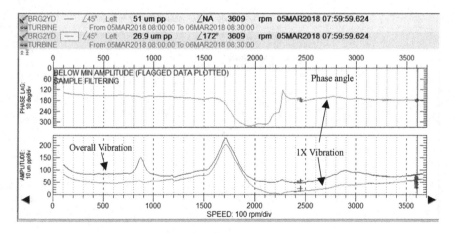

Fig. 2.16 Exemplary Bode plot

The Bode plot shown in Fig. 2.16 indicates the amplitude and phase by two separate graphs, whereby the horizontal axes are the speed and the two vertical axes are the amplitude and the phase angle, respectively.

The upper amplitude and phase graph represent the overall vibrations. There is no frequency, therefore, no phase angle. The lower amplitude and phase graph represent the rotational frequency component (unbalance).

2.3.3 The Polar Plot

In the polar plot, the amplitude and phase are integrated into one plot, whereby the plot represents the speed or time trace of the end point of the vibration vector.

In the variable speed plot, a "loop" represents an amplitude and phase change at the same time and therefore a critical speed or a resonance.

Figure 2.17 represents a polar plot with speed stamps. One can also change to time stamps. Figure 2.18 shows the same polar plot, but with time stamps.

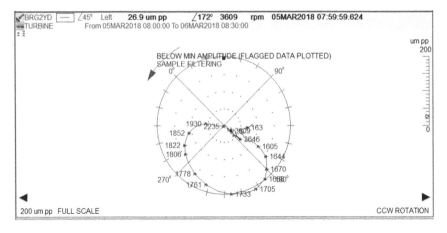

Fig. 2.17 The polar plot with speed stamps

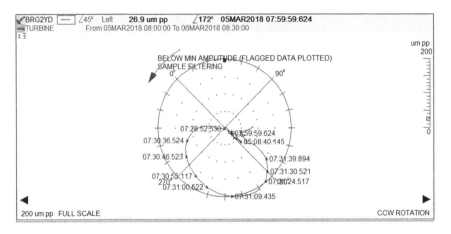

Fig. 2.18 The polar plot with time stamps

2.3.4 The Shaft Centerline Plot

The shaft centerline plot visualizes the gap voltage of both eddy current sensors A and B to the shaft at the same time, to obtain a graph of the distance or its change between the eddy current sensors and the shaft (see Sect. 2.1.2). That can be a function of time or a function of speed. The development of the oil film during speed change can be seen, but also quasi-static movements between shaft and pedestal at constant speed.

The shaft centerline plot (see Fig. 2.19) is used to judge the functionality of a journal bearing.

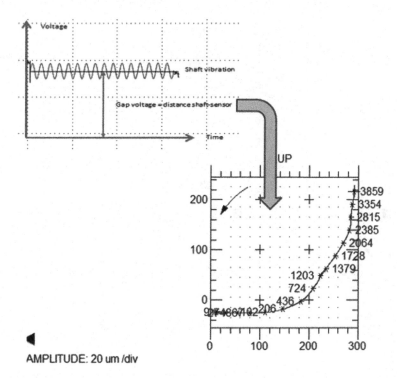

AMPLITUDE: 20 um /div

Fig. 2.19 Exemplary shaft centerline plot

2.3.5   *The Orbit/Time-Base Plot*

The orbit/time-base plot shows the kinetic movement of the shaft. Since the shaft transducer signals are relative vibrations between pedestal and shaft, also pedestal vibrations can contribute to the orbit/time-base plot. The orbit/time-base plot (see Fig. 2.20) is composed of the signals of the two eddy current sensors, located 90° to each other and in the correct timing sequence.

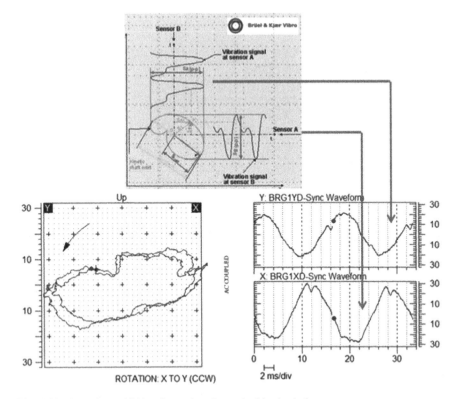

Fig. 2.20 Exemplary orbit/time-base plot: shows the kinetic shaft movement

2.3.6   The Waterfall Spectrum Plots

The waterfall spectrum plot is a 3-dimensional spectrum having time/speed, amplitude and frequency as parameters.

It is very useful for the determination of events besides an unbalance. This can be instabilities, cracks, resonances, events where other vibration frequencies are expected than the rotation component 1X.

The waterfall spectrum plot in Figs. 2.21 shows the run-up of a steam turbine, where there were intermissions at constant speed of 1100 and 2500 rpm.

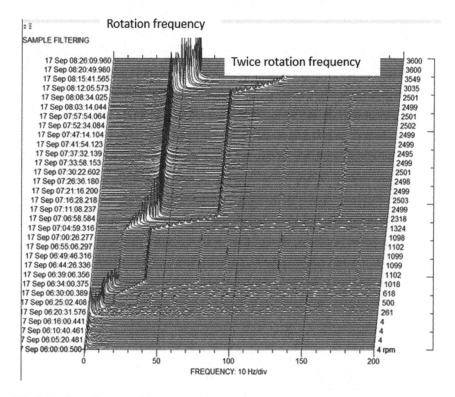

Fig. 2.21 Exemplary waterfall spectrum plot

Chapter 3
Fault Analysis: Vibration Causes and Case Studies

This chapter contains a collection of various case studies from 1969 up to now, covering almost every vibration event having taken place in power plants worldwide.

Vibration events are covered by this book or this chapter in particular:

Vibration causes:

- unbalances (1X vibrations): mass–thermal–magnetic,
- instable (non-repeatable) unbalance,
- non-homogenous rotor material,
- unusual GT starting behavior,
- instable (non-repeatable) unbalance,
- structural resonance problems,
- mechanical looseness,
- instant vibration increase (step change),
- alignment faults—coupling errors,
- rubbing,
- spiral vibrations—rotating vectors,
- 2X vibrations by sag,
- magnetic 2X vibrations,
- lateral rotor cracks,
- ring-shaped rotor cracks,
- instabilities,
- oil film in bearings, media flow and
- gas turbine compressor surge, rotating stall.

Table 3.1 covers the vibration problems to which we had been engaged during the past 50 years. To classify the problems, the best first diagnosis criterion (according to our judgment) will be the vibration frequency. Based on frequency, we divided the vibration problems into two main families:

© Springer Nature Switzerland AG 2020
F. Herz and R. Nordmann, *Vibrations of Power Plant Machines*,
https://doi.org/10.1007/978-3-030-37344-3_3

i. Forced vibrations: Vibration frequencies or multiples related to the rotation frequency fn
ii. Self-excited vibrations: Vibration frequencies not related to the rotation frequency fn.

Every one of the mentioned items in these two "problem families" will be covered with one or more case studies. The following pages will cover these case studies including the problem explanations and the countermeasures derived from these explanations.

Some of our problem explanations might not be strictly in line with some theoretical interpretations, but since the countermeasures based on these explanations were successful, the reasoning proved not to be wrong and these explanations might probably help to improve the theoretical calculation models.

Table 3.1 Different vibration causes

Forced vibrations: *Frequency of vibration f corresponds to rotation frequency fn or multiples 2 to 3 fn*		Self-excited vibrations: *Frequency of vibration f has no integer relation to rotation frequency*	
Caused by:		Caused by:	
1.	Excessive unbalance $f = fn$	1.	Excitation by the oil film in the bearing with $f \neq fn$, usually $f = fcr1$
2.	If $f = fn$: critical speed too close to operating speed If $f \neq fn$: usually $f = fcr1$ or flow of driving medium	2.	Excitation induced by steam
3.	Resonance of stationary part too close to operating speed, $f = fn$	3.	Excitation by internal friction, $f \neq fn$, usually $f = fcr1$
4.	Faults in coupling and alignment, $f = fn$ and $2fn$, possible harmonics	*Legend*: f: = frequency of vibration fn: = frequency of rotation $fcr1$: = 1st critical speed	
5.	Rotation of vectors, $f = fn$		
6.	Vibration due to magnetic forces, $f = fn$ and $f = 2fn$		
7.	Vibration excited by sag, $f = 2fn$		
8.	Rotor with lateral cracking		

3.1 Mass Unbalance

3.1.1 Rigid Rotors

Here, under the (allowed) assumption that the structure behaves like a linear spring, the vibrations reflect the centrifugal force. The rotor operates within the rigid speed range, which means that its critical speed will be far above its operating speed. The transient vibrations (run-up, run-down) follow solely the centrifugal force increase by speed.

The transient vibrations (run-up, run-down) follow the centrifugal force F_m, as a "basic curve," which is more or less a square function according to the development of centrifugal force with rotation speed ω. The kinetic activity of rotor (and support system, critical speeds, resonances) will be superimposed over this curve.

The rotational (unbalance) vibrations shown in Fig. 3.1 follow solely the centrifugal force.

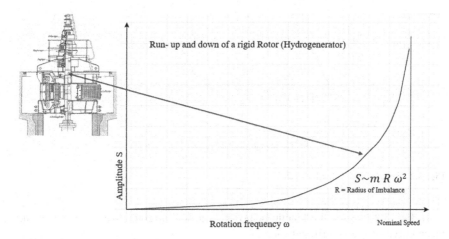

Run- up and down of a rigid Rotor (Hydrogenerator)

$S \sim m\,R\,\omega^2$

R = Radius of Imbalance

Amplitude S

Rotation frequency ω

Nominal Speed

Fig. 3.1 Behavior of a rigid rotor

The horizontal axis represents the rotor speed, at the vertical axis is the vibration deflection S. S follows approximately the centrifugal force.

3.1.2   Flexible Rotors

Also, here we might notice this "basic curve" as well (dependent on the unbalance amount and distribution), but as soon as the rotor passed its 1st critical speed, it changes from a "solid block" into an "elastic rod" and starts to deflect, governed by the unbalance distribution.

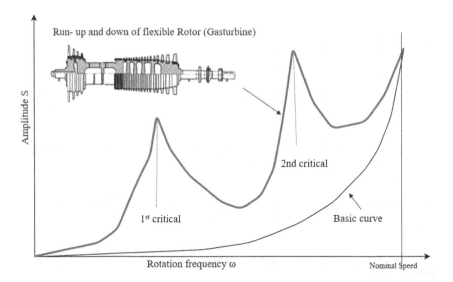

Fig. 3.2 Behavior of a flexible rotor

In Fig. 3.2, we see the kinetic behavior of a flexible rotor, which overrides the basic curve of the centrifugal force.

We can state for every one of these two cases:

If there is a progressive increase in the vibrations with operation speed, we most likely have a mass unbalance to be corrected by further balancing.

3.2 Thermal Unbalance

3.2.1 Example of a 120 MW Gas Turbine

This is an example of a 120 MW gas turbine in Holland. Slight irregularities in the blading, most likely at the compressor (like jamming blade roots at one side of the circumference and more looseness on the other side) cause tensions as soon as the engine starts and the temperatures at the rotor surface begin to rise.

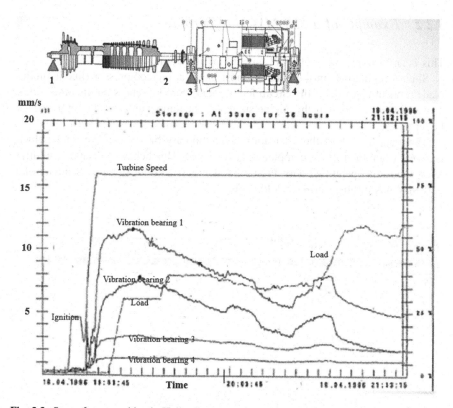

Fig. 3.3 Start of a gas turbine in Holland

Figure 3.3 shows how the vibrations of the initially cold turbine follow the temperature which increases with the load. After the ignition of the turbine, every load increase causes a vibration increase, but in general the vibrations tend to decrease until the rotor is heat soaked. The consequence of mentioned tensions is a temporary bending of the rotor, which recovers when the rotor is completely heat soaked. This example shows this typical behavior.

At ignition, we see a sharp vibration increase (only at the two gas turbine bearings), mostly exceeding the trip level, then a slow decrease follows, indicating the vibration recovery with increasing rotor core temperature. Every load increase is accompanied by a recovering vibration excursion. After approx. 10 h, the vibrations have a normal level again, when the rotor is heat-soaked.

Experience showed that this effect can age away after a certain operation time (respectively a certain number of heating cycles by starting). The countermeasure at jobsite will be a balancing exercise, whereby the added weight could be removed later, when the thermal imbalance faded away through ageing.

3.2.2 Example of a 250 MW Gas Turbine

This is an example of a 250 MW gas turbine in Spain.

Slight irregularities in the blading, most likely at the compressor (like jamming blade roots at one side of the circumference and more looseness on the other side), cause tensions as soon as the engine starts and the temperatures at the rotor surface begin to rise.

The consequence of these tensions is a temporary bending of the rotor, which recovers when the rotor is completely heat soaked. This behavior disappeared after a few thousand operating hours. Figure 3.4 demonstrates the vibration behavior of a 250 MW gas turbine during a cold start.

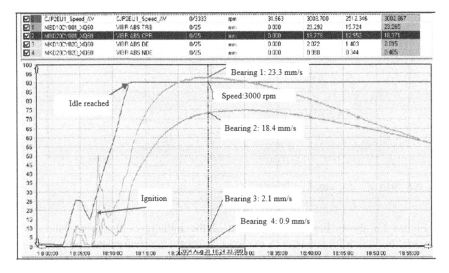

Fig. 3.4 Cold start of a 250 MW gas turbine

3.2.3 450 MW Combined Cycle Plant, Compromise Balancing

This is a typical example of a temporary thermal imbalance at a 450 MW turbo-set. It consists of the gas turbine (1), the fixed coupled generator (2), the self-synchronizing SSS clutch (3) and the steam turbine (4), which drive the generator as well but as soon as deliver torque at nominal speed (see Fig. 3.5). Since the vibrations during commissioning had been over trip level, a "compromise balancing" exercise (finding a good compromise between hot and cold condition) had been carried out.

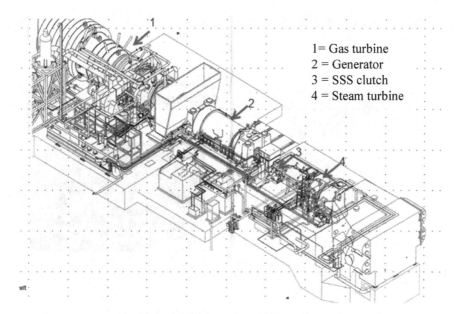

Fig. 3.5 Single-shaft combined cycle plant

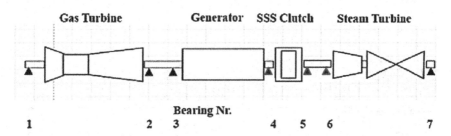

Fig. 3.6 Bearing arrangement

Figures 3.5 and 3.6 show a single-shaft combined cycle plant with steam and gas turbine on the same shaft. At first, the gas turbine will be started and delivers waste heat for the boiler. As soon as the boiler is on temperature, it delivers steam for the steam turbine. The steam turbine will then be started. When at rated speed, torque can flow to the generator, the SSS (Speed Self Synchronizing) clutch engages and the generator is driven from both ends (by the gas turbine and also by the steam turbine) and delivers now 450 MW full load.

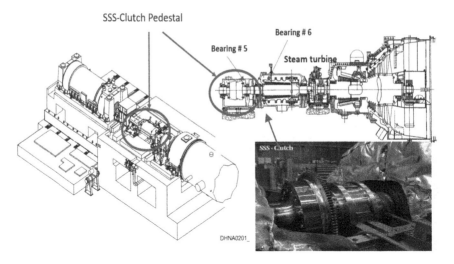

Fig. 3.7 SSS clutch pedestal

In Fig. 3.7 we see the arrangement of the SSS clutch pedestal.

When the GT is running up at cold, we see a fair vibration behavior at both bearings (green curve of Figs. 3.8 and 3.9).

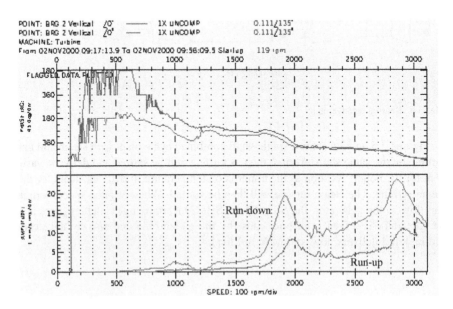

Fig. 3.8 Run-up (green) and run-down (red) before balancing bearing #2

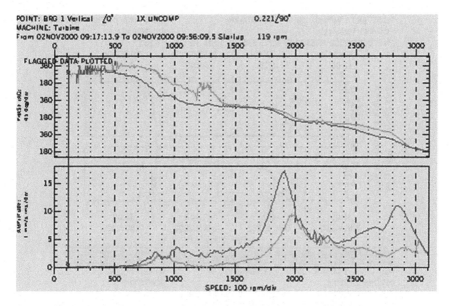

POINT: BRG 1 Vertical /0°   1X UNCOMP   0.221/90°
MACHINE: Turbine
From 02NOV2000 09:17:13.9 To 02NOV2000 09:56:09.5 Startup   119 rpm

Fig. 3.9 Run-up (green) and run-down (red) before balancing bearing #1

The horizontal axes of Figs. 3.8 and 3.9 represent the speed in rpm. The vertical axes of the upper charts show the phase, and the vertical axes of lower charts show the amplitude of the vibrations.

At rated speed, the vibration levels start to increase and after operating for approx. 30 min, they exceeded trip level and the machine had to be shut down due the thermal bending of the rotor. Another elevation of vibrations can also be seen passing the 2nd critical speed (at around 2850 rpm).

The red curves in Figs. 3.8 and 3.9 show the vibrations during run-down. When the 1st critical speed is passed (at around 1800 rpm), the vibrations are elevated, due the thermal bending of the rotor. Another elevation of vibrations can also be seen passing the 2nd critical speed (at around 2850 rpm).

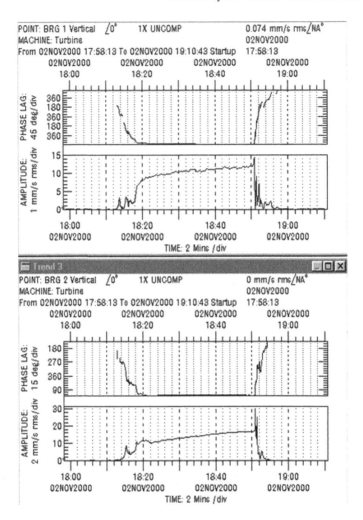

Fig. 3.10 Time trend bearing #1 and #2

In the time trend Fig. 3.10, we see a fair run-up in cold condition and the subsequent vibration rise. After about 30 min of running time at rated speed and idle, we initiated shut down followed by a relatively rough run-down.

Since the vibration rise takes place quickly with a short time constant, we can conclude that the thermal effect leading to vibrations takes place on the surface of the rotor.

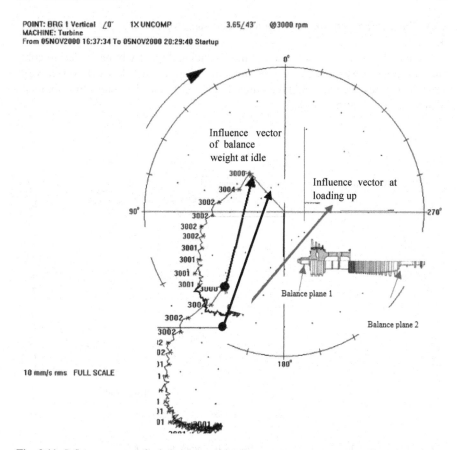

POINT: BRG 1 Vertical ∠0°   1X UNCOMP   3.65∠43°   @3000 rpm
MACHINE: Turbine
From 05NOV2000 16:37:34 To 05NOV2000 20:29:40 Startup

Fig. 3.11 Influence vectors for balancing and loading

The polar plot in Fig. 3.11 illustrates the compromise balance action of a gas turbine plant in Chile at turbine bearing #1 (bearing #2 is behaving quite similar). The operating condition at the balance action was idle.

Before balancing, the thermal excursion at bearing #1 (and #2) exceeded the trip limit by far. The weight caused the thermal changing range to be within the vibration limits, but the thermal change itself had not been affected. At base load, when the rotor was heat soaked, the vibration change recovered.

A characteristic feature of a thermal unbalance is an in-phase development of the vibrations at both bearings. You will see the vibrations of the whole speed range being elevated, especially at the 1st critical (the thermal bow of the rotor matches perfectly to the 1st modal deformation). If relative shaft vibration is measured, you will also see an elevated mechanical run-out at low speed in the hot condition.

In this case, the balance weight for compensating thermal unbalance was 2400 grams into turbine-plane 1. During the following time (approx. 1 year), this weight had been removed (part by part) as the thermal unbalance disappeared by the increasing number of starts (heat cycles).

Figure 3.11 demonstrates the meaning of compromise balancing, which means shifting of the thermal changing range.

At the run-up and run-down after balancing, we see that the behavior has changed.

We now have, as a consequence of the balancing, higher vibrations during cold run-up (upper curve) and lower vibrations during warm run-down (lower curve), but in both cases, we are within the limits. Figures 3.12 and 3.13 show the vibration changes at transient conditions.

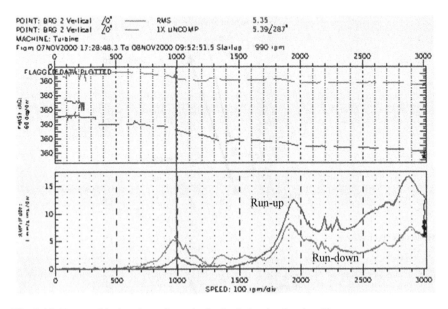

Fig. 3.12 Run-up (blue) and run-down (red) after balancing bearing #2

Here, we see the Bode plot of the compressor bearing #2. The horizontal axis represents the speed in rpm. The vertical axis of the upper chart shows the phase, and the vertical axis of lower chart shows the amplitude of the vibrations. The consequence of the balancing was the fact that the vibrations on bearing #2 at 1st critical speed (at 1800 rpm) are now elevated in the cold condition (run-up, blue curve). On the other hand, the vibrations decrease over the whole speed range, when the rotor becomes warm (run-down, red curve).

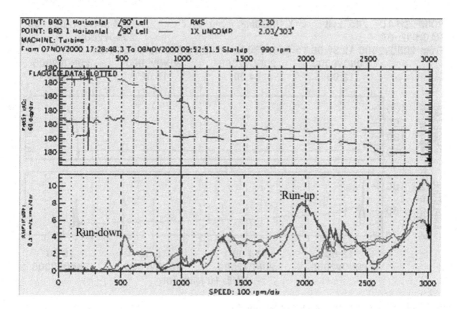

POINT: BRG 1 Horizontal /90° Left ——— RMS 2.30
POINT: BRG 1 Horizontal /90° Left ——— 1X UNCOMP 2.03/303°
MACHINE: Turbine
From 07NOV2000 17:28:48.3 To 08NOV2000 09:52:51.5 Startup 990 rpm

Fig. 3.13 Run-up (blue) and run-down (red) after balancing bearing #1

Here, we see the Bode plot of the turbine bearing #1. The horizontal axis represents the speed in rpm. The vertical axis of the upper chart shows the phase, and the vertical axis of lower chart shows the amplitude of the vibrations. The consequence of the balancing was the fact that the vibrations on bearing #1 at 1st critical speed (at 1800 rpm) are now elevated in the cold condition (run-up, blue curve). On the other hand, the vibrations decrease over the whole speed range, when the rotor becomes warm (run-down, red curve).

The compromise balancing changed the time trend as well and it shows a reversed behavior compared to the former trend. Now we start at higher vibrations and during the heat soaking process the vibrations decrease.

Figures 3.14 and 3.15, in contrary to Fig. 3.10, show dropping vibrations by time as a consequence of the balance weight.

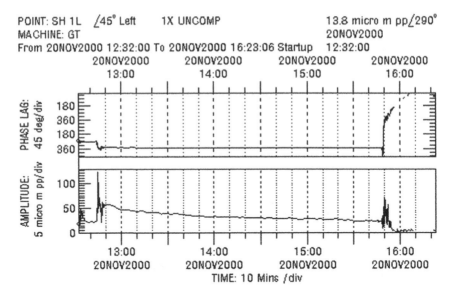

Fig. 3.14 Time trend after balancing bearing #1

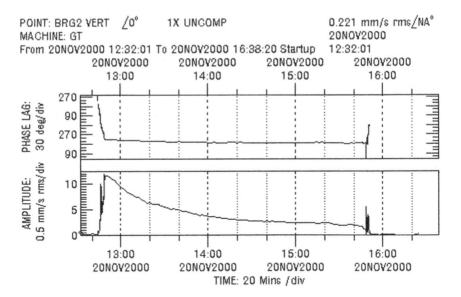

Fig. 3.15 Time trend after balancing bearing #2

3.2.4 Instant Vibration Increase (Step Change)

In the power plant in Sumatra, we have an example of an instant vibration increase happening during the startup of a gas turbine turbo-set. At the exciter rotor (overhang design) of the generator, a cooling element of the diode wheel moved due to poor assembly two times when nominal speed had been approached. Figure 3.16 shows the exciter of this 100 MW generator.

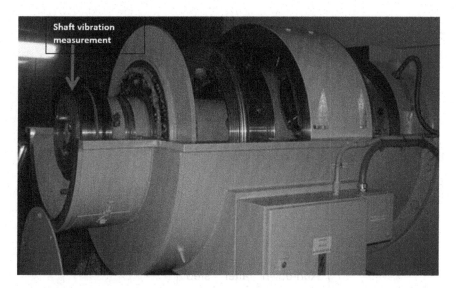

Fig. 3.16 Exciter of a 100 MW generator

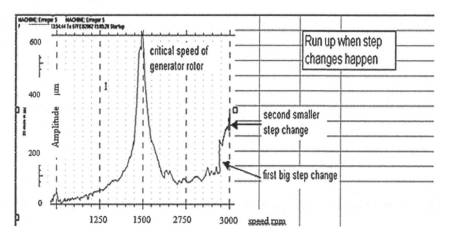

Fig. 3.17 Run-up at exciter

Originally, the exciter rotor was balanced so very well, that the exciter critical was not noticeable.

Figure 3.17 shows the run-up at the exciter. The horizontal axis shows the speed in rpm, and the vertical axis shows the amplitude of the vibrations. No exciter critical is visible, but two step changes close to the rated speed can be witnessed (see black arrows in Fig. 3.17 and enlarged detail in Fig. 3.18). We see a very pronounced 1st generator rotor critical at 1500 rpm, but no overhang critical which we would expect close to 3000 rpm.

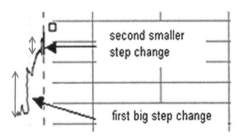

Fig. 3.18 Enlarged detail of vibration step changes

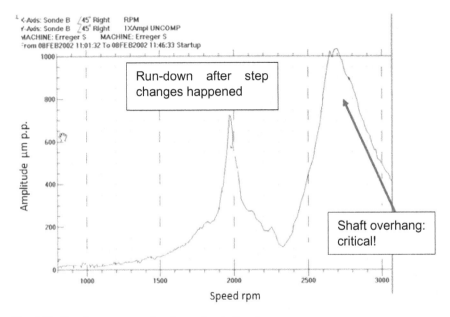

Fig. 3.19 Run-down, measured at the exciter shaft end

After the step change vibrations have happened (see Figs. 3.17 and 3.18), a dangerously high vibration of over 1000 μp_p. had been measured (relative shaft at outer slip ring) during run-down of the exciter critical speed (see Fig. 3.19).

After having fixed the cooling elements on the diode wheel and rebalancing, the step changes disappeared.

3.3 Non-homogenous Rotor Material

This a 200 MW steam turbine in Ireland. Here, the problem was in the intermediate pressure turbine.

A plant nearby Dublin had two of this turbo-sets—one was working perfectly, but the other one had this vibration problem since both machines had been converted from base load to peak load machines. The turbo-sets are started once a day and running for a few hours. About unit 2 (that with the problem), the client told us that it was possible to balance the rotor, but after a certain time with a couple of starts and stops, the balance condition worsened again, and it was necessary to rebalance again and again.

Figures 3.20 and 3.21 show the 200 MW the Irish turbo-set as a photograph and an assembly drawing.

Fig. 3.20 200 MW turbo-set

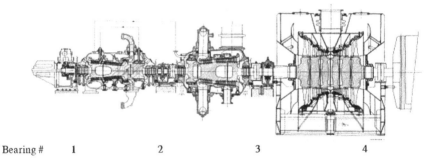

Bearing #   1   2   3   4

Fig. 3.21 Bearing arrangement

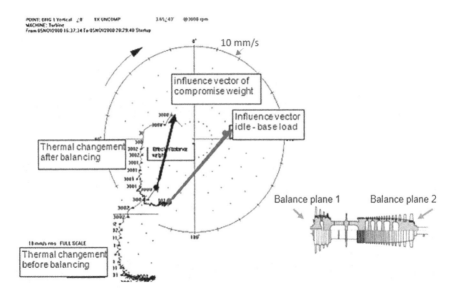

Fig. 3.22 Amrein surveillance chart

Figure 3.22 shows a turbine surveillance chart from the Amrein company that had been used in the 1950s for turbine supervisory. The horizontal axis shows time, and the vertical axis shows different parameters (temperatures, vibrations, etc.).

In Fig. 3.22, it can be seen that the vibrations of bearing #3 are developing almost synchronous to the inlet steam temperature development of the IP turbine (red arrows).

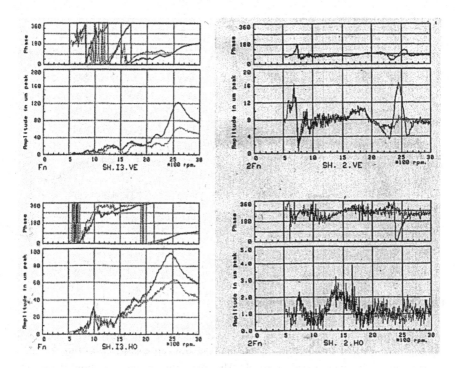

Fig. 3.23 Run-down warm and immediate run-up

Figure 3.23 shows the warm run-down and an immediate run-up.

From the shaft vibration readings, we had been drawing the deformation line, which is composed from the difference between run-up and run-down at the 1st LP critical, whereby every vertical line represents a measuring plane.

It indicated the expected thermal bending originating from the IP turbine rotor (see Fig. 3.24).

As a first approach to the problem, we also tried to rebalance the rotor, but it was not possible, because almost half of the circumference of the balancing planes at both sides (HP and LP side) had been already filled with balance weights.

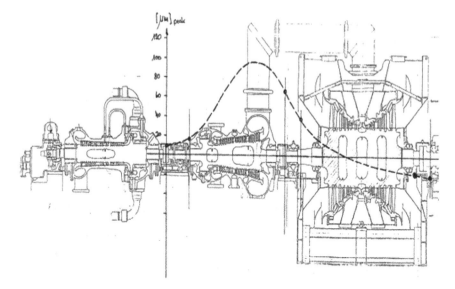

Fig. 3.24 Thermally induced rotor deflection

The turbo-set had been cooled down and the rotor had been taken out on a turning lath in order to measure the mechanical run-out according to Fig. 3.25.

The run-out showed a strange result: The maximum value was about 0.25 mm, which was almost a factor of 10 higher than allowed. The axial distribution of the run-out showed a kink in the center of the rotor, which could be interpreted as a lateral crack.

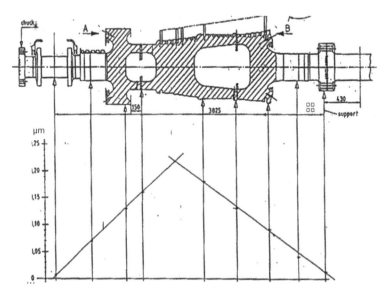

Fig. 3.25 Run-out on turning lath

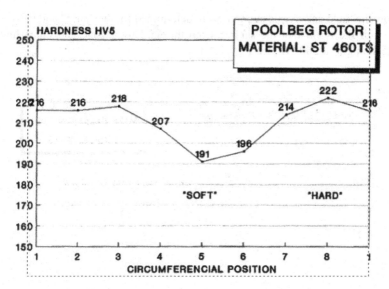

Fig. 3.26 Vickers hardness test

At first, a run-out test was carried out which showed a pronounced high point just opposite of the applied balance weights. The run-out did not change when the rotor was de-bladed; therefore, we suspected the reason for the bend being in the rotor body.

Very extensive ultrasonic testing did not give any indication for a crack which was a surprise for us and we continued by performing a hardness test. We found that the rotor hardness was not uniform, and a soft zone was located just opposite of the run-out high point (see Fig. 3.26).

Now the yield and tensile strength were determined for the "hard" and the "soft" zone, with the result that the sample from the hard rotor zone had much higher tensile strength at 500 °C (see Fig. 3.27).

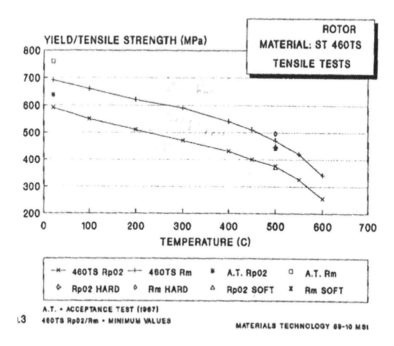

Fig. 3.27 Determination of the yield/tensile strength

ISO stress tests with 170 MPa showed that the hard zone had a much higher time to rupture within a temperature range of 500–600 °C (see Fig. 3.28).

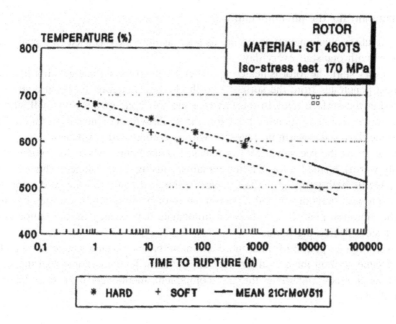

Fig. 3.28 ISO stress test

Finally, strain tests at operating temperature (590 °C) and a load of 170 MPa indicated a higher strain of the soft zone after a relative short time (6 h) and as longer load was applied, this difference in strain grew progressively (see Fig. 3.29).

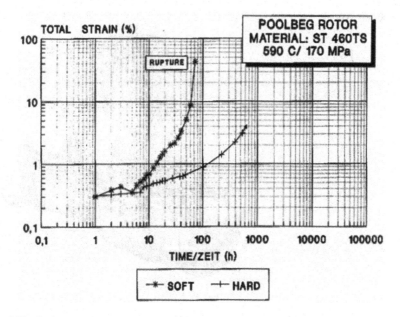

Fig. 3.29 Strain test

Explanation for the phenomenon:

This vibration phenomenon appeared, as soon as the unit was used as a peak load machine with one start every day.

At every start, the rotor is heated up, whereby the rotor surface gets hot first and the heat penetrates from the surface into the body. Therefore, a pressure stress is created at the surface. Due to the fact that the soft zone of the rotor will start to creep earlier and more than the hard zone, material will get a negative strain in the soft zone during the warm up phase: it will be compressed permanently.

As soon as the pressure stress is released (after hours, when the rotor is uniformly warm), a small rotor bending remains, causing an unbalance. This bending will have its high point in the hard zone. At the next start with the cold rotor, the whole process starts again, and a permanent rotor bending will grow step by step.

The vibration picture will show an immediate dependency on the temperature with a slow decreasing trend.

Over a longer period of time, more and more balance weights have to be added at the same spot on the circumference. The different hardness zones had the same effect as a crack regarding the 2X component mentioned later (see later in Sect. 3.13).

3.4   80 MW Gas Turbine Starting Behavior

The 80 MW gas turbine shown in Fig. 3.30 has an operating speed of 6212 rpm.

Fig. 3.30 80 MW gas turbine

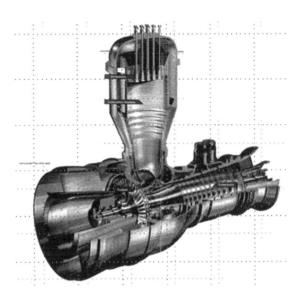

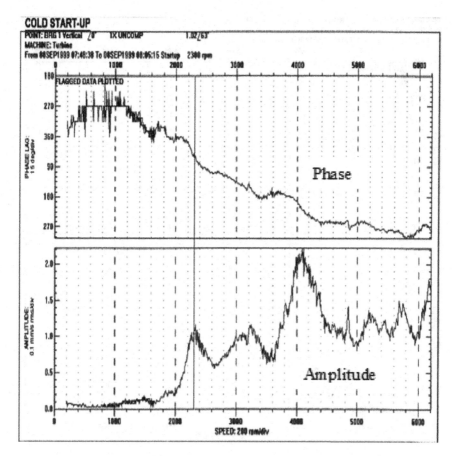

Fig. 3.31 Cold start of the gas turbine

In Fig. 3.31, a very fair cold start-up is displayed. The horizontal axis represents the speed in rpm. The vertical axis of the upper chart shows the phase, and the axis of lower chart shows the vibrations. In the lower chart, we see the 1st critical at about 2200 rpm and the 2nd critical at about 4100 rpm (elevated points).

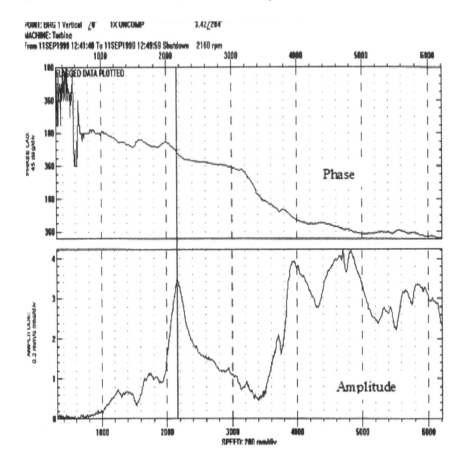

Fig. 3.32 Immediate hot run-down

Figure 3.32 shows an immediate run-down after a failed hot start attempt. The horizontal axis represents the speed in rpm. The vertical axis of the upper chart shows the phase, and the vertical axis of lower chart shows the amplitude of the vibrations.

The elevated vibrations at the hot start prompted us to try another run-up in a cooler condition. The hot start followed by the hot run-down (the engine went down to standstill, was probably 15–20 min on barring gear before start) showed extremely elevated vibrations when passing the 1st critical speed at approx. 2200 rpm (see Fig. 3.32).

So, we decided to cool the machine more down, by operating it on barring gear.

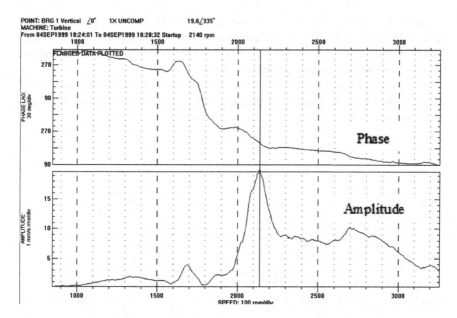

Fig. 3.33 Hot restart after 2 h

A restart after 2 h showed even higher vibrations than before (see Fig. 3.33), but only at the 1st critical around 2200 rpm. At the 2nd critical around 4000 rpm, we did not see a substantial change.

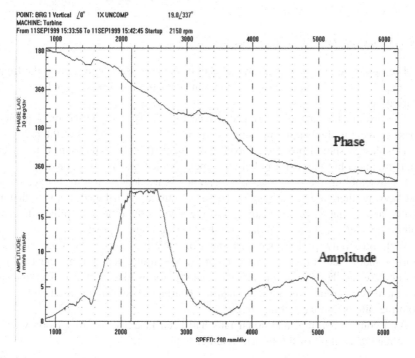

Fig. 3.34 Restart after 4 h

After 4 h downtime at barring gear, the situation worsened again (see Fig. 3.34). Almost 20 mm/s at the 1st critical has been reached, without influencing the 2nd one. It needed approx. 6 h of down time, to be able to start the engine again after a hot run-down without risking a vibration trip.

We found that the GT-rotor developed a bend while it was cooling down and therefore the 1st critical had been primarily excited. But that was not the whole story, because it did not explain the heavy vibrations entirely.

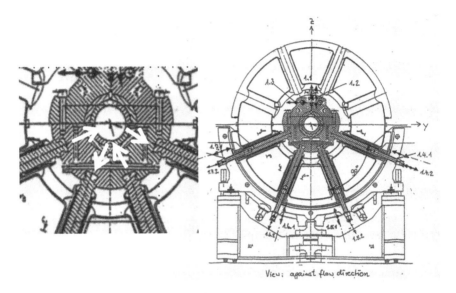

Fig. 3.35 Design of front bearing pedestal

The right side of Fig. 3.35 shows a detail of the cross-section design of the front bearing. The left side indicates an enlargement of the 4 supporting pillars of the bearing head.

The bearing housing of bearing one (see Fig. 3.35) is supported on 4 pillars. These connect to the bearing housing by means of keys in keyways of the housing.

This design should provide an axial movement (thermal expansion of the pedestal structure) but prevent a radial movement (which means vibrations).

For assembling reasons, a certain clearance is needed; therefore, between key and keyway, there are 0.2 mm gaps. These 0.2 mm are distributed in a way that at two pillars the gap is on the bearing housing side, at the other two pillars the gap will be opposite, and on the pillar side the bearing housing is clamped.

With proximity probes, we measured the relative motion between support bearing saddle and pillar attachment points. We found that during the cooling process, the saddle will not stay circular, and it deforms like an ellipse. That means that there is a certain time period when the bearing housing is released and loose from the pillars and can vibrate freely.

As a countermeasure, additional fans had been installed in order to provide a more regular cooling of the bearing saddle.

3.5 Magnetic Unbalance of a 20 MW Hydroplant in Switzerland

In this case study, we are dealing with a rotor of so-called rigid behavior (see Sects. 3.1.1 and 3.1.2). After a routine revision, heavy vibrations occurred at this remote-controlled power plant.

Fig. 3.36 Remote-controlled hydrostation

This is a small hydropower plant in the French part of Switzerland (20 MW). It consists of a horizontal Francis turbine and a 20 MW Generator and is used as a base load unit, running at full load the entire year (see Fig. 3.36).

There are normally no personnel in this plant, and the operation is controlled from a central station probably 100 km away. This is a unit with an 8-pole generator; therefore, the rated speed is 750 rpm.

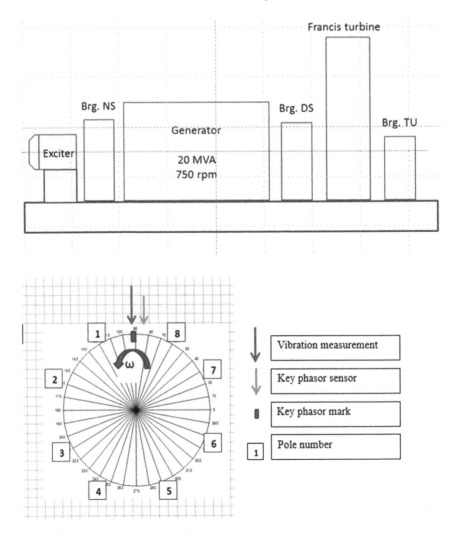

Fig. 3.37 Setup for measurement and balancing

Figure 3.37 shows the setup of the vibration measurement equipment, which was temporarily used to resolve the problem. As a measuring instrument, a Schenk Vibroport 40 had been used. The problem was that especially the non-driven side bearing showed heavy vibrations, which were coming instantly when the rotor excitation was switched on. The vibrations increased more when the load was raised to full load.

This was a new behavior which occurred after a big overhaul of the unit. Since it was an instant vibration change depending on the rotor excitation current with no time delay, we already suspected a magnetic unbalance. The solution was a compromise balancing exercise. Since the unit operates practically 100% at full load, the full load vibrations had been reduced while the idle vibrations increased as the consequence.

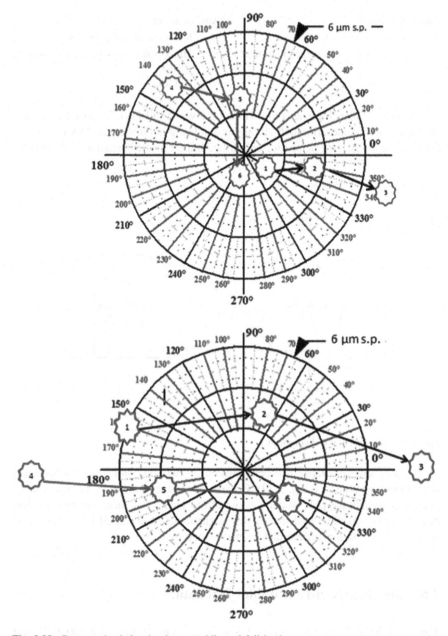

Fig. 3.38 Compromise balancing between idle and full load

Figure 3.38 shows the influence vectors of load and excitation current on the vibrations as a polar plot, before and after a compromise balance exercise.

Before balancing:		After balancing:	
1:	Idle operation, rotor unexcited	4:	Idle operation, rotor unexcited
2:	Idle operation, rotor excited to nominal voltage	5:	Idle operation, rotor excited to nominal voltage
3:	Full load	6:	Full load

The aim of balancing was a low vibration level at full load, because it is a base load machine. Temporarily high vibrations at idle did not disturb, and they occurred only during the starting period for only a few minutes. The balancing took place in that regard.

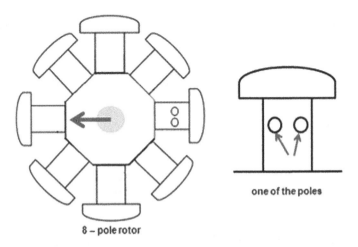

8 – pole rotor

one of the poles

Fig. 3.39 Poles of 20 MW hydroplant

The magnetic unbalance had been created by two axial holes, which had been drilled into one pole (see Fig. 3.39). Because of the fact that during the reassembly after the overhaul the pole wheel was out of mechanical balance, one side was too heavy. The two axial holes disturbed the magnetic field and caused 1X vibrations as soon as the excitation current had been switched on.

3.6 Instable (Non-repeatable) Unbalance

This is a 600 steam turbo-set in the USA. There were 2 units, each 600 MW, and the rated speed is 3600 rpm. The problem was an excessive vibration increase at the slip-ring shaft bearings #6 and #7.

It was not possible to operate the unit longer than 5–10 min, as the vibrations reached values that would jeopardize the rotor to disintegrate. Figures 3.40 and 3.41 show an example of a 600 steam turbo-set.

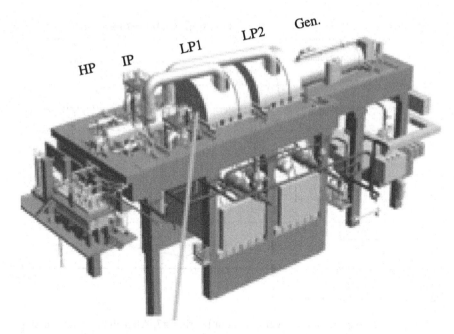

Fig. 3.40 600 MW turbo-set in the United States

This is a model of this turbo-set. The arrangement of the elements of the turbo-set is indicated below in Fig. 3.41.

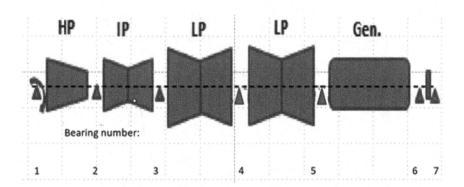

Fig. 3.41 Turbine and bearing setup

Figure 3.42 shows the redrawn surveillance chart, where the excessive vibrations can be seen:

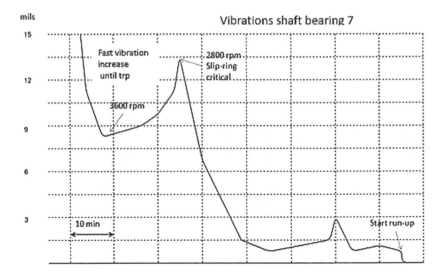

Fig. 3.42 Warm start of the unit

First, we see the vibration development at bearing #7. After that, we are passing the slip-ring critical at 2800 rpm with 13 mils shaft vibration and we reach 3600 rpm with about 8 mils. Then the vibrations increase very fast and within 10 min, we have to shut down, because the measuring range of 15 mils had been exceeded and there was no control about the vibration levels anymore at bearing #7. Bearing #6 showed at the same time 8–9 mils, when the unit was shut down.

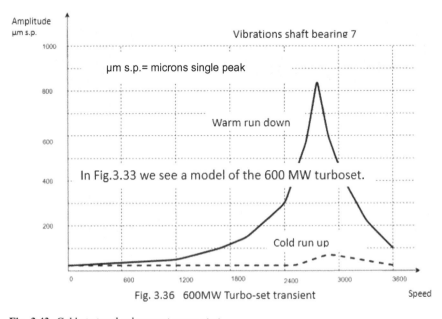

Fig. 3.36   600MW Turbo-set transient

Fig. 3.43 Cold start and subsequent warm start

Figure 3.43 shows the redrawn cold start and the subsequent warm run-down. There is a remarkable difference of vibrations at the slip-ring critical, which can only be explained by the increased temperature of the slip-ring fan (shrink fit looser).

Almost 800 $\mu m_{s.p.}$ had been reached, which corresponds to approx. 30 mils. Meanwhile in the cold run-up, we only measured hardly 100 $\mu m_{s.p.}$ (approx. 4 mils). To find out the root cause for this problem, we removed all the standing parts from the slip-ring shaft. We built a framework around the shaft to fix proximity probes for measuring the shaft vibrations and recorded them on a magnetic tape due to the short running time possible (Fig. 3.44).

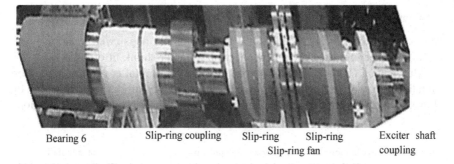

| Bearing 6 | Slip-ring coupling | Slip-ring | Slip-ring | Exciter shaft |
| | | Slip-ring fan | | coupling |

Fig. 3.44 Slip-ring shaft

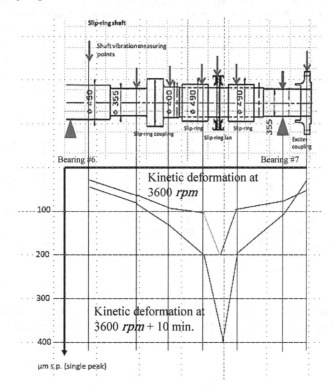

Fig. 3.45 Shaft measuring positions and measurement

Here, we see the different elements of a slip-ring shaft of a large generator rotor. Figure 3.45 documents the kinetic deformation.

After about 10 min at 3600 rpm excessive vibrations occurred, whereby an operational mode shape of the slip-ring shaft had been determined. But on top of this modal deformation, we saw a relative motion of the slip-ring fan in regard to the shaft, on which it was shrunk on.

Conclusion:

Loose shrink fit due to temperature and centrifugal force at rated speed caused the non-repeatable and excessive vibrations.

Solution:

The original fan between bearings #6, #7 and the slip rings (with a shrink fit of 0.8‰) had been replaced by a new one with a tighter shrink fit of 1.3‰.

After having changed the fan, we had a repeatable and (after rebalancing) a fair vibration behavior of the unit with max. shaft vibrations of 2–3 mils. Figure 3.46 shows the reading after the fan exchange at 400 MW in the heat-soaked condition.

BE = Bearing cap vibrations, SH= Shaft vibrations in μm p.p.   Steady state

Speed :	3600 rpm.		Status:	OK	Meas. Pt.			30 Jun 1988		17:06:00
Meas. Pt.	Fn		2Fn			Fn		2Fn		
	Ampl.	Phase	Ampl.	Phase			Ampl.	Phase	Ampl.	Phase
BE.1 .R	1.6	238	0.8	209	BE.1 .L		1.9	231	0.8	180
BE.2 .R	0.3	180	0.2	180	BE.2 .L		7.2	253	1.1	224
BE.3 .R	8.8	228	0.5	189	BE.3 .L		6.6	67	0.5	180
BE.4 .R	14.7	220	1.7	285	BE.4 .L		19.8	61	2.3	161
BE.5 .R	35.2	270	2.2	279	BE.5 .L		49.3	141	4.4	258
BE.6A .R	39.1	269	5.8	338	BE.6A .L		41.6	97	7.3	298
BE.6B .R	0.8	306	3.4	334	BE.6B .L		0.9	120	1.1	174
BE.7A .R	5.5	255	2.5	251	BE.7A .L		3.4	133	6.9	88
BE.7B .R	9.7	238	26.7	286	BE.7B .L		11.2	187	28.3	105
					.L					
SH.1 .R	13.1	131	1.3	247	SH.1 .L		16.1	48	2.3	277
SH.2 .R	13.6	251	1.1	168	SH.2 .L		3.9	191	2.8	252
SH.3 .R	26.1	224	1.4	174	SH.3 .L		22.3	24	1.3	180
SH.4 .R	42.7	219	3.4	147	SH.4 .L		57.2	63	3.4	279
SH.5 .R	39.2	261	3.9	176	SH.5 .L		74.7	165	0.3	180
SH.6A .R	23.9	251	27.5	249	SH.6A .L		13.9	142	22.7	154
SH.6B .R	21.3	100	37.9	561	SH.6B .L		20.0	303	17.6	147
SH.7A .R	45.5	205	17.7	83	SH.7A .L		31.9	139	17.3	4
SH.7B .R	55.8	308	22.9	225	SH.7B .L		27.4	152	16.9	63
SH.SF .H	23.6	197	0.2	73			0.5	276	0.3	139

Fig. 3.46 Vibrations after repair

The vibrations with the tighter shrink fit of the fan showed a significant improvement of the slip-ring shaft vibrations at bearing 6, 7 and the fan itself.

3.7   Structural Resonance Problems

3.7.1   200 MW Steam Turbine, the Netherlands

This here is an elder example (1982) of a 200 MW steam turbine in the Netherlands where the first-time synthetic modification was used.

This is a technology which uses the computer model from modal analysis, to change the structural properties (masses, stiffnesses) virtually.

Figure 3.47 shows the design drawing of the huge steel foundation from MAN, which was standard in the Netherlands in the 1980s (there are some steel foundations in Holland like Fig. 3.47, because of the soft sand underground).

The steel foundation is over 50 m long and about 20 m high and has a weight of about 400 metric tons (which is much less than the weight of the turbo-set mounted on top of it). The problem bearing was bearing #4 between the low-pressure turbines.

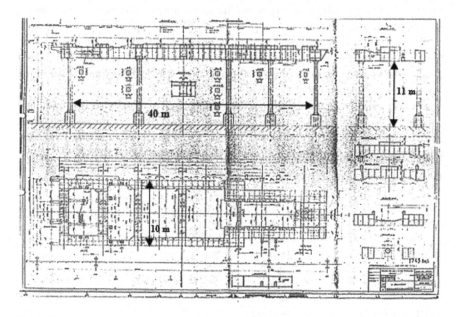

Fig. 3.47   MAN, Steel foundation of 200 MW turbo-set

The problem of this machine was caused by the change of the operation concept. It was changed from a base load to a peak load machine, which was started and stopped every day. It had a manual synchronization device; therefore, it took a long time to catch the right moment of switching the generator to the grid.

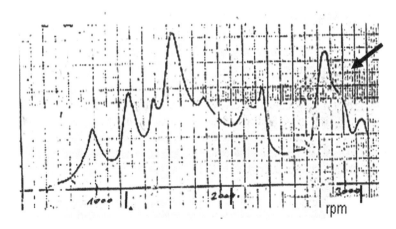

Fig. 3.48 Vibrations during run-up

Figure 3.48 shows a recording of an old X/Y recorder, which are not in use anymore. The horizontal axis represents the speed in rpm. The vertical axis shows the amplitude of the vibrations.

As can be seen on the run-up vibrations in Fig. 3.48 (which were measured in $\mu m_{s.p.}$ single peak), there is a resonance peak of bearing #4 in vertical direction close to the operating speed, whereby the operating speed (3000 rpm) is still in the slope of the resonance curve (black arrow).

Looking at the vibration readings table, there are two things conspicuous and typical for structural resonances: A structural resonance always shows a preferred vibration direction; in this case, it is vertical. Here, we see much higher vibrations than horizontally.

Table 3.2 Vibrations readings at 100 MW

Bearing and shaft vibration at 100 MW, all values in $\mu m_{s.p.}$		Pedestal vibrations	Absolute shaft vibrations
1. HP front bearing #1	Horizontal	40	–
	Vertical	42	–
2. Thrust bearing bearing #2	Horizontal	20	–
	Vertical	–	–
3. LP front bearing #3	Horizontal	–	–
	Vertical	8	–
LP1–LP2 bearing #4	Horizontal	7	7
	Vertical	**25**	**22**
Generator DS (driven side)	Horizontal	7	1
	Vertical	16	25
Generator NDS (non-driven side)	Horizontal	10	10
	Vertical	11	10
Auxiliary bearing	Horizontal	5	35
	Vertical	16	8

The significance of bold should highlight bearing 4 between low pressure 1 and low pressure 2 turbine as a bearing with almost the same high vibrations in thevertical direction, which is a clear indicator of a vertical pedestal resonance

Table 3.2 shows that the relationship of the absolute vibrations between shaft
and pedestal is small, which means little or no relative motion between shaft and
bearing in the resonating direction (here vertical).

The level of pedestal vibration is not only due to unbalance excitation, it is also
caused by the elevated mobility of the structure. A modal analysis had been carried
out with the structural model shown below (see Fig. 3.49). As an excitation force
for the analysis, an electro-hydraulic shaker system had been used. The two shaker
heads (max. 10 kN each) had been attached at the arrow positions. The shakers
operated in-phase and vertically. The shaker positions had been chosen, because we
suspected the highest unbalance input at the generator driving end.

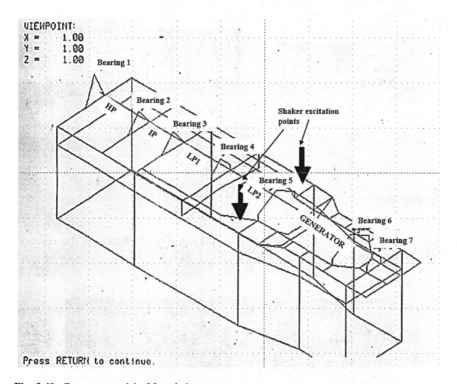

Fig. 3.49 Computer model of foundation

Figures 3.50, 3.51, 3.52 and 3.53 show the electro-hydraulic shaker system built
by Servo Consultants, UK, according to our specifications. It is capable of 10 kN
alternating force within 5 to 200 Hz.

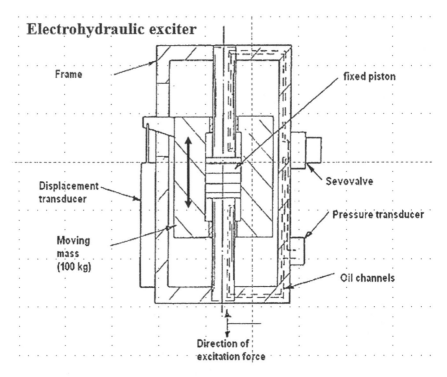

Electrohydraulic exciter

Frame

fixed piston

Displacement transducer

Sevovalve

Pressure transducer

Moving mass (100 kg)

Oil channels

Direction of excitation force

Fig. 3.50 Electrohydraulic exciter (shaker)

In Fig. 3.50, the shaker principle is displayed. The excitation is the reaction force of the moving mass.

Fig. 3.51 Photograph of the shaker head

Figure 3.51 displays the 110 kg shaker head with its high-pressure hoses. To provide the shakers with the necessary oil pressure of max. 5000 psi, there is a pump unit with an 85 hp driving motor. The high-pressure pump, driven by the motor is connected to the actuators by means of high-pressure hoses visible in Fig. 3.51.

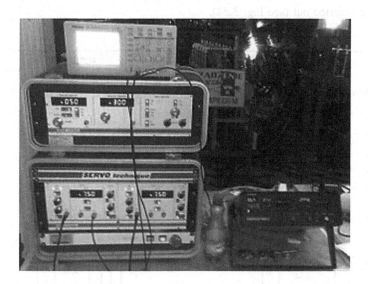

Fig. 3.52 Control unit of shaker system

Figure 3.52 shows the control unit of the system.

Fig. 3.53 Pump unit of the system

Comparing the transfer functions of the different bearings, we saw a pronounced vertical resonance close to 50 Hz at bearing #4. This was also confirmed by an impact test. The other bearings did not show any disturbing resonances close to the operating frequency.

To confirm the resonance behavior at bearing #4, impact tests of bearing #4 to #7 had been carried out (see Fig. 3.54).

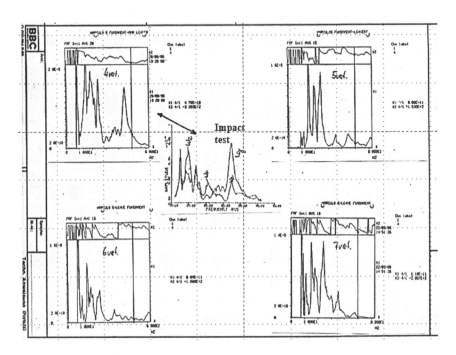

Fig. 3.54 Shaker and impact tests

The shaker tests at the bearings 4 to 7 vert. show a pronounced peak at bearing 4 vert. close to the operating speed at 45 Hz (see graph in upper left corner). Therefore, it had been verified with an impact (bump test), which is shown in the center of Fig. 3.54.

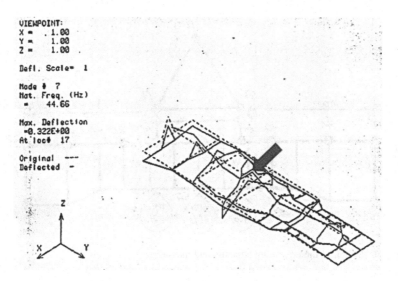

Fig. 3.55 Foundation mode shape close to operating speed

The modal analysis showed a 1st bending mode of the crossbeam on which bearing #4 is attached at 45 Hz (red arrow, Fig. 3.55).

The bearing pedestal is "riding" on the vertically vibrating cross-beam. One of the reasons of its high flexibility is the fact that there are no vertical pillars underneath the foundation table (see Fig. 3.49). There is no space for pillars because of the condenser underneath the LP cylinders. This was a general problem of the big MAN steel foundations. Our company counteracted this problem; therefore, continual thermally caused alignment changes with the "one bearing design" between the turbine stages and made the turbo-set insensitive to alignment changes and bearing instabilities.

The only remaining possibility to tune the beam resonance away from the operating speed was an attachment of additional mass, which brings the frequency down. To quantify this mass, we applied the synthetic modification. This is a feature of the more advanced modal analysis programs.

A kinetic model is established in the computer, which is based on the equation of motion and of course the FRFs measured at the respective measuring points. (An FRF stands for frequency response function and is a diagram of the kinetic stiffness in m/N over the frequency in HZ. It is obtained by exciting structures with an external, measured force). FRFs are consisting of a mass, damping and stiffness matrices having the resolution of the measuring point pattern. If you want to change a mass, you must go into the mass matrix and exchange the masses there (which are derived from the FRFs) to the desired values. Subsequently, a new resonance frequency will be calculated. We had been modeling the crossbeam according to the sketch below (see Fig. 3.56) and had been adding 30 tons at the indicated points underneath bearing #4. Figure 3.57 shows the result of the modification calculation.

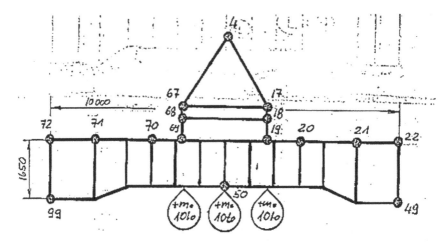

Fig. 3.56 Sketch of the crossbeam structure and measuring

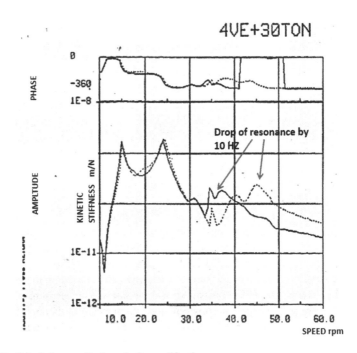

Fig. 3.57 Calculation result of synthetic modification

Afterward, we applied this additional mass physically. This was done by filling the crossbeams' center compartments gradually with sand and keeping on measuring. After having applied 23 tons of sand, we had already achieved the desired 10 Hz of resonance frequency down-shift (see Fig. 3.58).

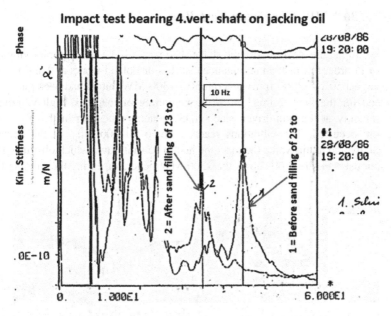

Fig. 3.58 Impact test after filling with sand

The run-up, run-down vibration plots showed now a resonance free region at rated speed (see Fig. 3.59). The beam resonance had been shifted down to 1950 rpm, which cured this problem.

After a trim balance correction, the vibration behavior of the turbo-set was perfect.

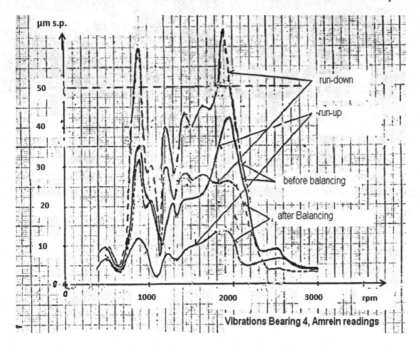

Fig. 3.59 Running test and balancing after filling with sand

3.7.2 1300 MW Generator Test Run

In this example, the electrohydraulic shaker had been used as a temporary vibration absorber in order to finish an acceptance test. In our Poland works at that time, one of the largest 50 Hz (3000 rpm) generators of 1300 MW output had been put on the test stand for the final testing. From the beginning of the tests, high vibrations, predominantly at the non-driven side (NDS) generator bearing in the horizontal direction, occurred. The vibrations reached up to 90–100 μm_{pp} at the pedestal horizontally and at this value the machine had to be shut down which obstructed the tests seriously. Figure 3.60 shows the 1300 MW generator during the running test.

Fig. 3.60 1300 MW Generator on the test foundation

A modal analysis had been carried out by sinusoidal excitation, whereby our electro-hydraulic shaker system had been used again. The two inertia actuators had been attached to the NDS generator bearing horizontally as shown in Fig. 3.61 (the shaker device is displayed and described on pages 2 and 2).

Fig. 3.61 Shaker head

Several attempts for improvement had been made without the desired improvement (see Fig. 3.62):

- attaching additional masses and
- mounting additional stiffening beams.

Therefore, we refrained from further measures, concerning mass or stiffness changes.

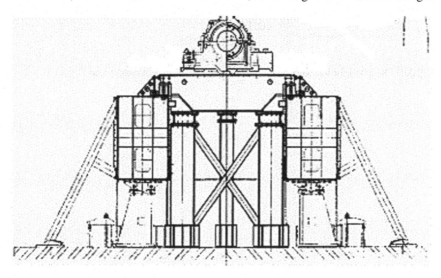

Fig. 3.62 Several non-effective attempts

After several attempts failed, the modal analysis had been carried out in the final condition of the test foundation concerning the reinforcement measures.

The transfer function (response) measured at the driving point (bearing NDS horizontally) is shown in Fig. 3.63.

We see a static flexibility (at frequency = 0) of 1.7 E−9 m/N as well as a pronounced horizontal resonance at 50.7 Hz, almost at operational speed. We assume that the static flexibility in the initial status of the structure was even higher, leading to the high vibrations horizontally. The additional measures reinforcement measures may have reduced the static flexibility, but unfortunately a horizontal resonance had been created almost exactly at operating speed, resulting again in a horizontal vibration problem.

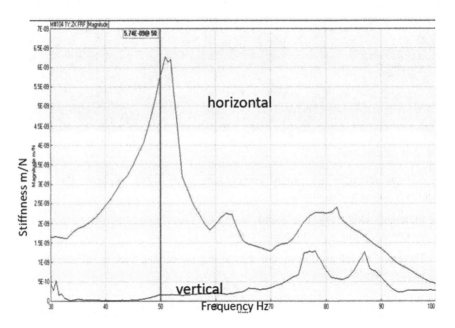

Fig. 3.63 Response of the NDS bearing—horizontal resonance at rated speed

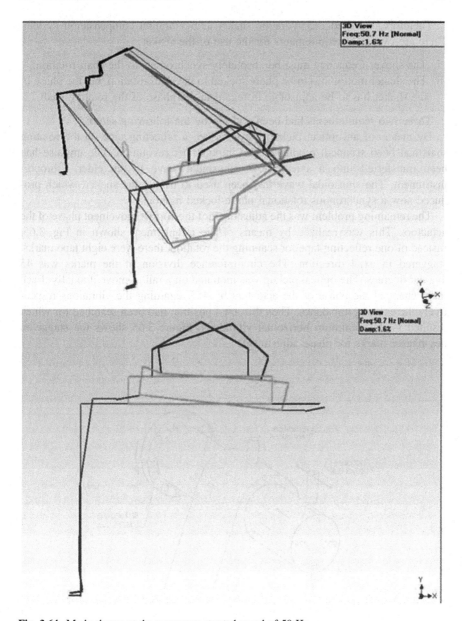

Fig. 3.64 Mode shapes at the resonance at rated speed of 50 Hz

Regarding the fact that this problem only occurs at the test installation and is expected not to occur at the much stiffer and heavier concrete foundation in the plant, we were thinking of an active damper being applied during the tests (Fig. 3.64).

Therefore, we used the hydraulic shaker as a vibration compensation system. There are two main requirements for the use of the shaker:

i. The shaker frequency must be absolutely synchronous to the shaft rotation.
ii. The shaker motion has to be phase-locked to the shaft rotation, but the phase of the shaker has to be adjustable in regard to the phase of the rotating shaft.

These two requirements had been realized by the following setup:

By means of an optical pickup (key-phasor), a reflecting tape on the rotating shaft had been scanned, producing one impulse per revolution. The impulse had been transferred into a synchronous sinusoidal wave by an elder Vibroport instrument. The sinusoidal wave had been used to trigger the shaker, which produced now a synchronous rotational phase-locked motion.

The remaining problem was the adjustment of the correct movement phase of the actuators. This was realized by means of the arrangement shown in Fig. 3.65. Instead of one reflecting tape for scanning the rotation, there were eight tape marks, staggered in axial direction. The circumference division of the marks was 45 angular degrees. The optical pickup was mounted on a rail to move it axially. Each mark changed the phase of the actuators by 45° regarding the vibrations respectively the exciting unbalance. Then this axial position had been searched for which resulted in the minimum horizontal vibrations. Figure 3.65 shows the staggered key-phasor marks for phase adjustment.

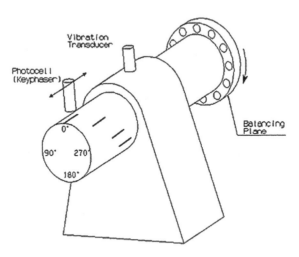

Fig. 3.65 The phase angle adjustment

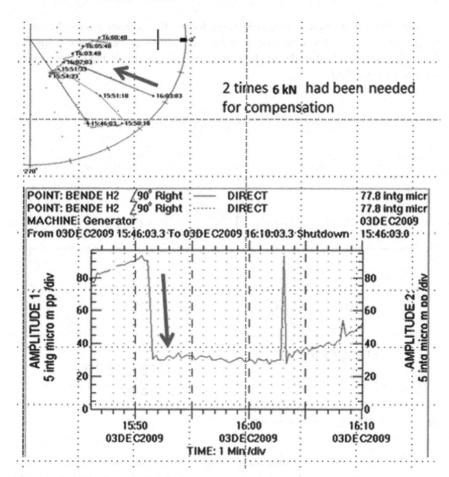

Fig. 3.66 Effect of shaker engagement

Figure 3.66 shows how the actuator decreases vibrations at rated speed. Here, we see the effect of the shaker regarding the rotor speed. The actuator engagement at rated speed is displayed as polar plot and time trend.

The red arrow indicates the vibration decrease in the polar plot and the time trend: At 15:50, the actuators had been engaged and the vibrations had been reduced from 90 to 30 μm_{pp}. At 16:03, the shaker had been disengaged and engaged again for control purposes. This caused a sudden increase in vibrations (see peak in time trend of Fig. 3.66).

Unlike the tuned absorber, the reducing effect of the actuator covers the entire
speed range (see Fig. 3.67), because an unbalance as well as the actuator motion is
phase-wise phase-locked to the rotating shaft. The dedication of the shaker device
resolved this problem.

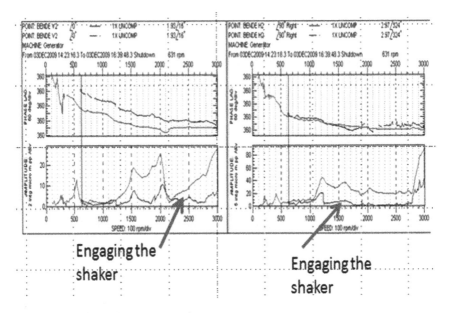

Fig. 3.67 700 MW turbo-set cross section

The red curve shows the vibration readings with a disengaged shaker, and the
black curve shows the readings with the engaged shaker. We see that the shaker
engaging influence improves the vibrations over the entire speed range, because the
shaker frequency and phase are controlled by the key-phasor signal. Therefore, the
shaker is phase-locked (Fig. 3.68).

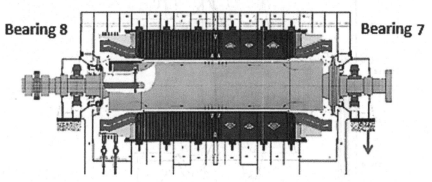

Fig. 3.68 700 MW turbo-set generator cross section

3.7.3 700 MW Turbo-Set in a Power Plant in Scotland

The protocol shown in Fig. 3.69, should prove that our newly retrofitted rotor has a crack. This was based on the following reasoning:

The rotor shows 3 critical speeds:

- 1st at around 1400 rpm.
- 2nd at around 2300 rpm and
- 3rd at 3000 rpm (which is the operating speed).

Looking at repeated measurements, the 2nd and the 3rd always change its frequency within a range of 200 Hz:

1. It is absolutely unrealistic, that a stiffness change due to the crack will result in a frequency changing range of 200 Hz.
2. A 1st critical at a large rotor like this, must be much lower than 1400 rpm.
3. The relation of the claimed critical speeds is completely unrealistic.

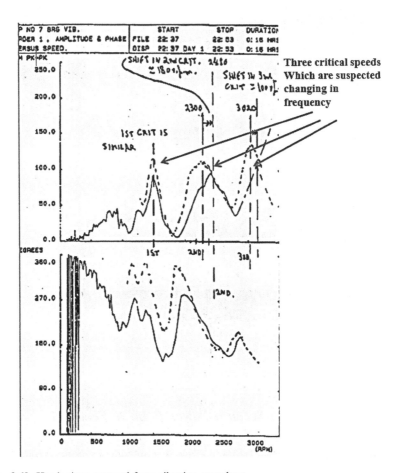

Fig. 3.69 Handwritten protocol from vibration consultant

A calculation had been made based on the rotor geometry and we found (see Fig. 3.70):

- 1st critical at 963 rpm and
- 2nd critical at 1461 rpm.

There is no other rotor critical below the operating speed. The consultant had identified the 2nd critical as the 1st one and the so-called 2nd and 3rd critical are no rotor critical speeds; they must be structural resonances.

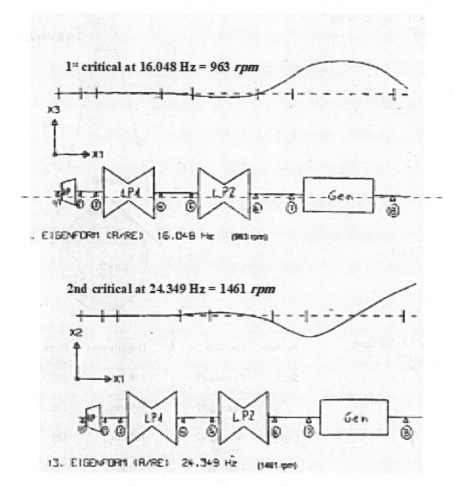

Fig. 3.70 Recalculated critical speeds and mode shapes

Figure 3.70 shows the mode shapes of the turbo-set for two different critical speeds. The critical speeds had been recalculated now showing the correct critical speeds. Since we already suspected a structural resonance problem, we tried to back that up with additional measurements. An effective method to identify structural resonances is the comparison between shaft and bearing cap vibrations.

We compared the absolute shaft vibrations measured with a "fishtail" (a simple softwood but effective shaft rider being pressed against the rotating shaft, see Fig. 3.71), with the absolute bearing cap vibrations.

At both generator bearings, we see a lower shaft than cap vibration, which indicates resonance behavior of the generator frame. Moreover, we see that shaft and the cap at bearing #8 (NDS generator) are vibrating antiphase. Here, the shaft, rotating in this bearing, will not be the vibration excitation. The bearing cap is here excited from a resonating frame.

Fig. 3.71 Principle of the "fishtail"

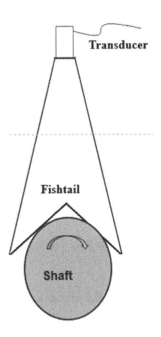

Table 3.3 Vibration readings

Position	Location	Displacement	Phase angle
Shaft	Bearing #7 vertical	108 μm_{pp}	202°
	Bearing #7 horizontal	110 μm_{pp}	135°
Cap	Bearing #7 vertical	162 μm_{pp}	189°
	Bearing #7 horizontal	160 μm_{pp}	148°
Shaft	Bearing #8 at 45°	55 μm_{pp}	194°
Cap	Bearing #8 at 45°	90 μm_{pp}	25°

Shaft absolute and cap vibrations in $\mu m_{p.p.}$ (Table 3.3):

Analyzing driven side (bearing #7) versus non-driven side (bearing #8):

- Shafts absolute vibrations are in phase.
- Caps absolute vibrations are in antiphase.

Therefore, the dynamic exciting force is at the driven side, bearing #7. Compared to the other similar units of the plant, this unit (unit no. 2) has a much higher mobility.

A vibration mapping around the generator fixation to the foundation showed a complete looseness of the generator. The foundation rail was loose on the foundation and the generator was loose on its foundation rail.

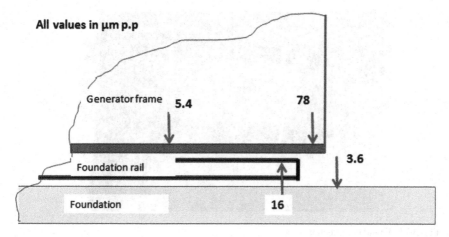

Fig. 3.72 Vibration mapping underneath the driven side of generator

The measured values displayed in Fig. 3.72 were obtained with the roving transducer connected to the vibration analyzer.

The conclusion of the mapping: The generator frame and the foundation rail are loose.

Underneath the generator, we saw that the alignment shims were coming out because of the relative motion of the supporting element between generator and foundation (see Fig. 3.73).

Fig. 3.73 Indicates the loose shims underneath the generator

The photograph of the generator bottom (see Fig. 3.73) shows the consequence of the "jumping" generator.

The reason for the vibration problem was a relatively thick layer of paper between the anchor nut of the foundation anchor bolt and the concrete. Therefore, the anchoring of the generator foundation was loose.

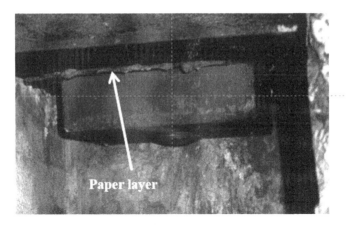

Fig. 3.74 Foundation anchor

This paper layer (see Fig. 3.74) had been forgotten from the 1st assembly.

After having corrected the attachment of the foundation to the concrete, this problem was resolved.

The successful foundation repair is documented in Fig. 3.75. As predicted, the last generator critical speed is the 2nd one around 1400 rpm. Also, the 1st one at around 900 rpm can be noticed.

Fig. 3.75 Measurements before and after repair

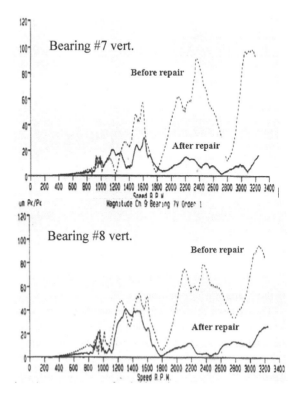

3.8 Axial Vibrations

3.8.1 300 MW Turbo-Set in Switzerland

The following picture (see Fig. 3.76) shows one of multiple 300 MW turbo-sets of a nuclear plant.

Fig. 3.76 300 MW Turbo-set

Figure 3.77 shows the bearing arrangement of the 300 MW plant steam turbine. There we had a resonance problem at bearing #3 between the two low-pressure turbines ND1 and ND2.

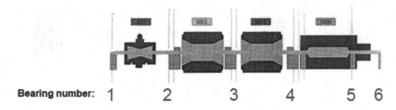

Fig. 3.77 Bearing arrangement

As can be seen in Fig. 3.78, bearing #3 showed an extremely high axial vibration.

All amplitude values in $\mu m_{p.p.}$

onstantbetrieb
Anzeige in um e.A.

Drehzahl: 3010 1/min. Status: ALARM				2 Sep 2005 10:47:40	
Meßstelle	Fn	2Fn	Meßstelle	Fn	2Fn
	Ampl. Phase	Ampl. Phase		Ampl. Phase	Ampl. Phase
LG.1. R	2.0 198	.8 123	LG.1. L	5.0 76	.2 218
LG.2.2 V	5.8 45	1.6 166	LG.2.2 H	12.0 210	.7 343
LG.3.1 V	9.3 147	1.4 23	LG.3.1 H	11.4 188	.6 136
LG.3.2 V	19.0 218	1.1 278	LG.3.2 H	12.2 196	.6 144
LG.4. V	5.0 290	6.9 290	LG.4. H	3.4 314	3.9 265
LG.5. V	5.5 140	2.8 335	LG.5. H	3.9 74	1.5 294
LG.6. V	2.5 269	.5 224	LG.6. H	5.8 237	.8 277
WE.1. R	6.2 276	1.8 20	WE.1. L	18.2 16	1.6 111
WE.2.2 V	7.1 328	3.2 114	WE.2.2 H	33.2 206	5.2 261
WE.3.1 V	17.1 198	1.5 228	WE.3.1 H	20.1 118	.9 45
WE.3.2 V	37.2 213	1.5 206	WE.3.2 H	35.4 184	2.3 284
WE.4. V	3.8 258	6.0 290	WE.4. H	20.7 70	4.6 206
WE.5. V	10.5 222	5.6 327	WE.5. H	26.5 142	7.0 215
WE.6. V	18.3 329	2.6 160	WE.6. H	39.9 234	4.8 346
EXZENTR.	4.7 80	.3 180	LG.2 AX	4.6 315	1.9 12
LG.3. AX	77.0 73	4.3 46			

Legend:

LG: Bearing (pedestal) vibrations

R: Right, V: Vertical, AX: Axial, H: Horizontal

WE: Relative shaft vibrations

Fig. 3.78 Vibration listing of surveillance system. Legend: *LG* Bearing (pedestal) vibrations, *R* Right, *V* Vertical, *AX* Axial, *H* Horizontal, *WE* Relative shaft vibrations

The run-down vibrations according to Fig. 3.79 unveil an axial resonance in the Bode plot of the bearing pedestal, which is practically corresponding with the nominal speed.

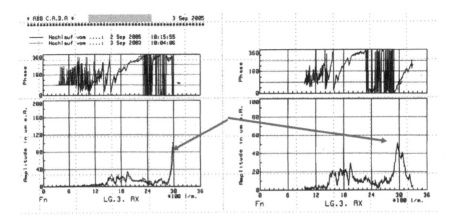

Fig. 3.79 Shaft vibrations during run-down

The kinetic movement of the pedestal—an axial rocking—can be very dangerous. It can cause a fracture of the supporting structure underneath the pedestal due
to material fatigue and this could lead to a destruction of the complete turbo-set. In
Fig. 3.80, the design of bearing #3 is displayed.

Fig. 3.80 Design and axial
movement of bearing #3

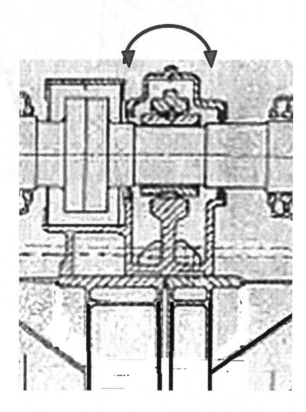

The solution with a detuning mass had been chosen. A steel block of 400 kg had
been attached rigidly on top of the bearing to lower the axial resonance frequency.
The additional masses had been attached as can be seen in Fig. 3.81.

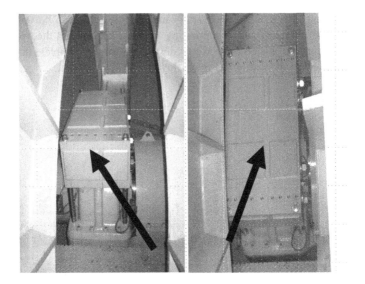

Fig. 3.81 Additional mass at pedestal 3

This mass lowered the resonance frequency from 3000 to 2800 rpm and the amplitudes at rated speed by a factor of 7–8 (see Fig. 3.82).

Generally, we can say the following:

If we can approximate the resonance system with an SDOF (single degree of freedom, only one mode), we will need 20% of the vibrating mass to achieve 10% frequency shift downward.

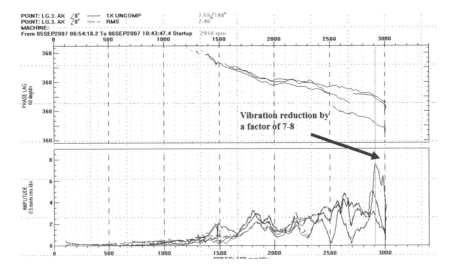

Fig. 3.82 Run-down after attaching the mass

3.8.2 Combined Cycle Plant in Saudi Arabia

This case is about a steam turbine shaft line of a multi-shaft combined cycle (CC) plant. There are three bearings B1–B3. Each bearing has two radial relative shaft proximity probes (45° offset) and a vertical absolute velocity transducer on the bearing cap. There were no sensors in axial direction. Figure 3.83 shows the bearing arrangement of the steam turbo-set.

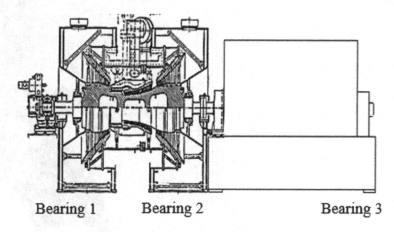

Bearing 1   Bearing 2   Bearing 3

Fig. 3.83 Turbo-set and bearing arrangement

The vibrations measured during transient and stationary operating conditions were very fair. There was no indication of a problem, which could be attributed to vibrations (see Fig. 3.84).

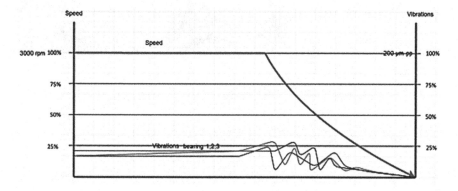

Fig. 3.84 Vibration chart: no indication to vibration problems

Despite the low vibrations indicated in Fig. 3.84, the jacking oil lines broke at
both ends of the generator, which could be dangerous because the escaping oil
could catch fire.

At the arrow position (see Fig. 3.85), cracks already had been discovered at the
jacking oil lines and investigated (see Fig. 3.86). At a first glance, the cracks looked
like caused by vibrations.

Fig. 3.85 Jacking oil line 1

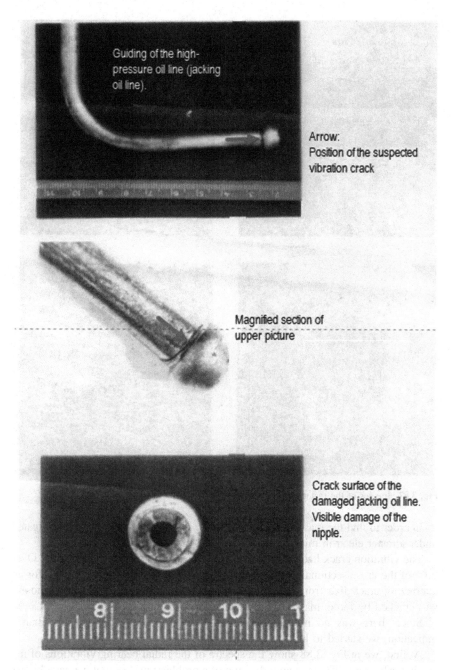

Fig. 3.86 Vibration crack confirmed at jacking oil line

The crack had been investigated closer with an electronic microscope (see Fig. 3.87).

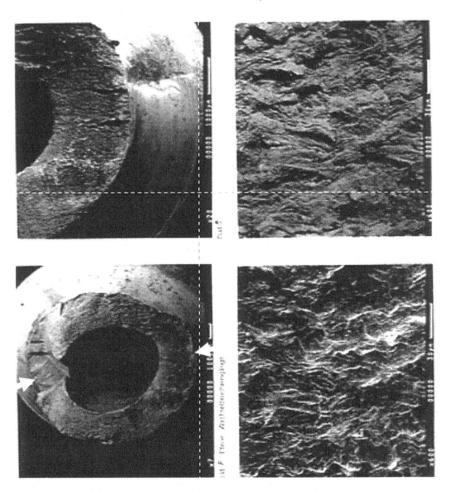

Fig. 3.87 Images of the raster electron microscope

In order to find the reason for the cracks, the crack surface had been investigated under a raster electron microscope.

The vibration crack had been confirmed by means of microscope imaging. Over 90% of the cross-sectional area looked like "water waves," which is typical for an alternating crack like from vibrations. A small remaining area, which is smooth, was cracked by force, marked by the white arrows.

Since there was no indication of high radial vibrations by the surveillance indication, we started to look at the axial vibrations.

At first, we in Fig. 3.88 show the spectra of the radial bearing vibrations of the generator bearings. There is no indication of a problem; we see that we have higher vibrations in the horizontal direction (max. 3.4 mm/s), but the vibrations are monitored vertically. We also see that especially at bearing #3, we have a considerable 2X component.

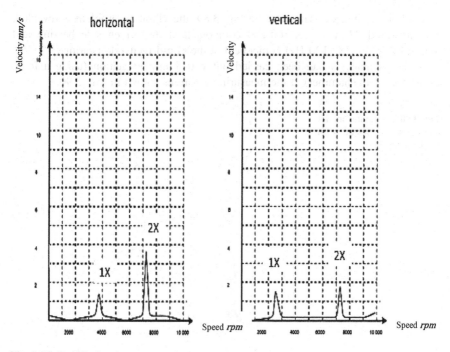

Fig. 3.88 Radial vibrations

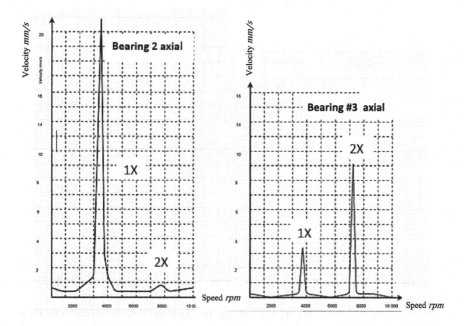

Fig. 3.89 Axial vibrations of bearing #2 and #3

Looking at the axial vibrations in Fig. 3.89, the vibration problem is revealed. We measured 22 mm/s 1X (60 Hz) component at the driven side bearing and almost 10 mm/s 2X (120 Hz) component at the non-driven side bearing.

A modal analysis had been carried out with bump excitation (using a force measuring sledgehammer) at the measuring positions shown in Fig. 3.90.

Fig. 3.90 Computer model and measuring positions for modal analysis

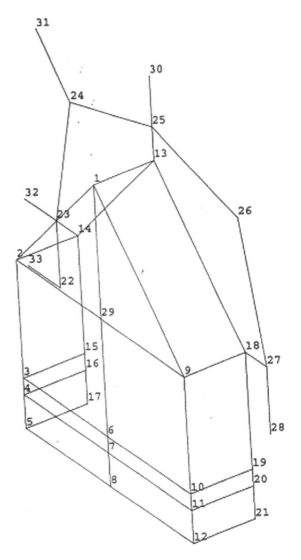

It showed resonance behavior of both bearing pedestals at "forbidden" frequencies.

In Fig. 3.91, we see a rocking motion with 60 Hz at the driven end and a rotation with 120 Hz around the vertical axis.

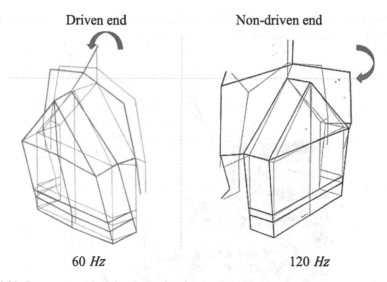

Driven end   Non-driven end

60 *Hz*   120 *Hz*

Fig. 3.91 Resonances at both bearing pedestals

These mode shapes shown in Fig. 3.91 had been confirmed also during operation according to the vibration mapping measurements.

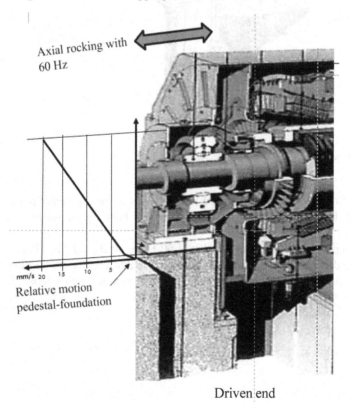

Axial rocking with
60 Hz

mm/s 20   15   10   5

Relative motion
pedestal-foundation

Driven end

Fig. 3.92 DS pedestal movement in operation

At 60 Hz, we measured the maximum vibrations at the top center position of the pedestal. The pedestal performed an axial rocking motion.

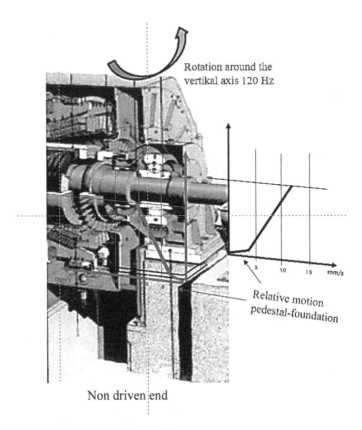

Fig. 3.93 NDS pedestal movement in operation

At 120 Hz, we measured almost no vibrations at the top center position of the pedestal, because the measuring point there had been on the rotation axis of the mode.

An important detail is the fact that at 60 Hz as well as at 120 Hz at the driven and non-driven side, we discovered a considerable relative motion between the pedestal foundation plate and the foundation due to a flexible (or loose) foundation connection (Figs. 3.92 and 3.93).

Fig. 3.94 Temporary stiffening bars

To prove the resonance theory, a temporary fix according to Fig. 3.94 had been attached.

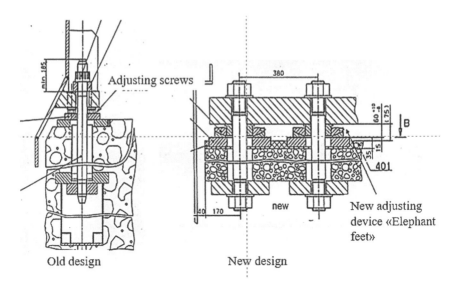

Fig. 3.95 Improvement of the foundation connection

The new design called "elephant feet" (Fig. 3.95) should replace the old adjusting screws.

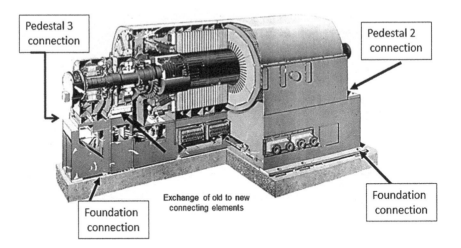

Fig. 3.96 Spots of the foundation connection

Final solution for the NDS bearing pedestal resonance problem was the following.

The "old" adjusting screw foundation connections, which could not prevent a relative motion between pedestal and foundation (especially at 120 Hz), had been replaced at the spots marked in Fig. 3.96 by so-called elephant feet connections (Fig. 3.95).

These elephant feet had been introduced firstly under the NDS pedestal and finally at all the indicated positions according to Fig. 3.96.

After improving measures

			Vibrations mm/s	
			Dec. 95	After improving measures and balancing
Bearing	2 axial	60 Hz	15	. 6,6
		120 Hz	4	0,6
Bearing	3 axial	60 Hz	3.7	3,3
		120 Hz	9.5	1,9

Fig. 3.97 The final vibration values

Finally, the rotor was successfully trim-balanced, and the problem was resolved according to the final vibration measurement Fig. 3.97.

3.8.3   150 MW Steam Turbine at the Philippines

In Fig. 3.98, we see the turbo-set at the Philippines and its bearing arrangement in Fig. 3.99.

Fig. 3.98 Here we see the 150 MW turbo-set

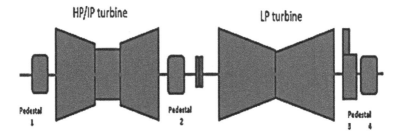

Fig. 3.99 Shaft setup and bearing arrangement

At this unit, excessive axial vibrations had been discovered by chance, because there were no axial surveillance transducers installed.

The high axial vibrations occurred at pedestal #4 between the LP turbine and the generator (see Fig. 3.100). It could be assumed that there was an axial resonance close to operating speed (3600 rpm).

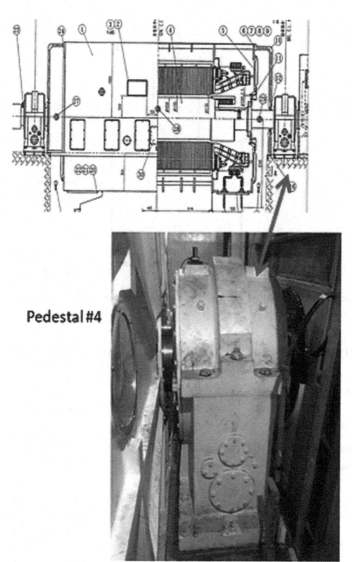

Fig. 3.100 Vibrating pedestal #4

We decided to do a vibration mapping (roving a transducer from measuring position to measuring position). This will help to find the origin of the high vibrations (Fig. 3.101).

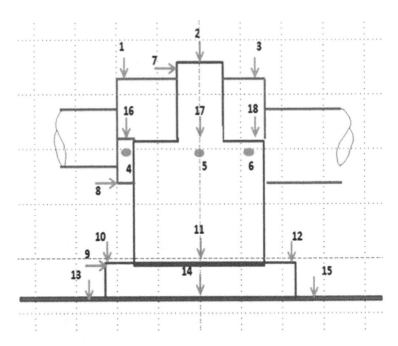

Fig. 3.101 Measuring points of the vibration mapping (see also Fig. 3.100)

Table 3.4 Measuring values of vibration mapping

Position	mm/s	mm/s	Position	mm/s	mm/s
All values in $\frac{mm}{s}$ r.m.s					
Conversion displacement to velocity: $\mu m_{p.p.} = \frac{\frac{mm}{s} r.m.s}{Hz} \times 450$					
A: without stiffeners, B: with stiffeners					
	A	B		A	B
1	8.4	6.5	10	4.5	0.5
2	1.3	1.3	11	4.9	1.4
3	7.3	2.9	12	4.4	3.1
4	2.3	1.9	13	3.8	1.7
5	3.1	2.1	14	1.6	1.6
6	3.1	2.4	15	3.5	2.8
7	**31.7**	**11.1**	16	5.8	0.7
8	**20.5**	**5.8**	17	2.2	2.4
9	2.8	2.3	18	5.7	3.6
Pedestal 3 axial	4.1	3.5			

Bold indicates the highest, most critical readings

Here is a "vibration mapping" of pedestal #4, and it can be noticed that there are extremely high vibrations (in condition A) in axial direction. The axial direction is not integrated in the surveillance system (because the ISO Standard does not require it). Nevertheless, these high values must be corrected otherwise material fatigue and damage could be the consequence. Condition B had been achieved by installing stiffeners between pedestal #3 and #4.

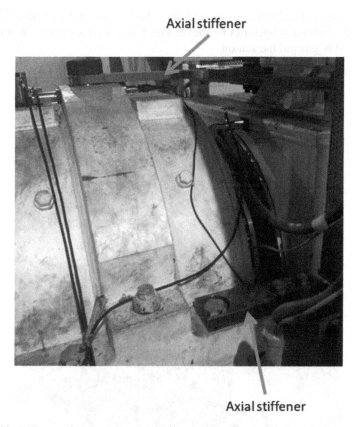

Fig. 3.102 Stiffeners of pedestal 4

The axial stiffening bars shown in Fig. 3.102 cured the problem.

The axial vibrations had been reduced by a factor of 3 not effecting the other vibrations (see results in Table 3.4).

3.9   Structural Resonance Problems

3.9.1   Vertical Machines

Resonances of vertical machines often occur at the big cool water pumps. There are
the bearing inspection openings in line. In Fig. 3.103, we see the cool water pump
of a 400 MW gas turbine station.

Fig. 3.103 Vertical cool water

If the central intermediate piece with the pump will be rotated by 90° regarding
the pump, the service openings will have a 90° offset. The supporting structure
would see a different stiffness in the relevant direction and the resonance might be
detuned.

3.9.2   Mechanical Looseness

If there is looseness in the structure, rattling will be the consequence. Rattling noise is a sequence of hits or impacts.

Figure 3.104 shows the motor of a cooling tower fan of a 400 MW gas turbine plant.

Fig. 3.104 Small squirrel cage motor

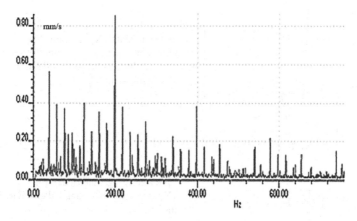

Fig. 3.105 Spectrum of looseness

Impacts (e.g., rattling) produce harmonic frequencies. Therefore, a spectrum indicating harmonics (especially at a point, where elements are connected) will be a feature of looseness. Figure 3.105 shows an example for such a spectrum of a loose foundation screw of a motor.

3.10   Alignment Faults—Coupling Errors

Figure 3.106 summarizes different coupling errors:

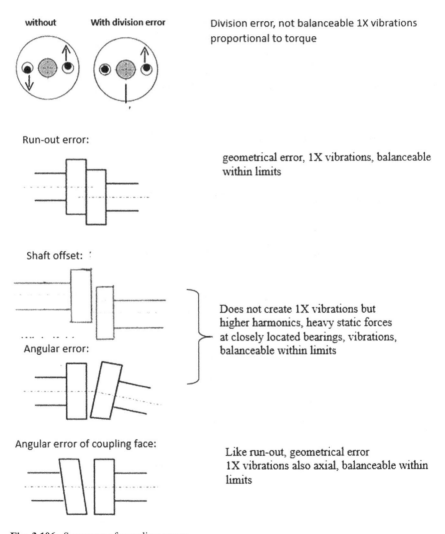

Fig. 3.106 Summary of coupling errors

Coupling errors will produce vibrations (with rotational frequency) in the following 3 cases:

- division error,
- radial run-out error and
- angular error of coupling face by wrong machining.

Angular misalignment and a shaft offset will not produce vibrations with rotation frequency. They will produce heavy static forces in the neighboring bearings. Run-out and coupling face errors occur nowadays seldom, but we saw a few division errors in the past.

3.10.1   100 MW Gas Turbine in Brazil

The intermediate block of the gas turbo-set displayed in Fig. 3.107 contains the SSS clutch and the gear of the side starter gear. The vibration problem was at the intermediate block.

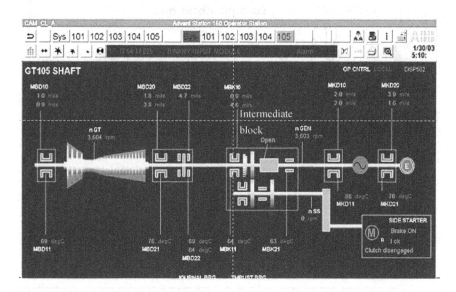

Fig. 3.107 Shaft system of the gas turbo-set

The relative shaft vibrations reached values close to trip level (200 $\mu m_{p.p.}$) at base load. When the unit is brought from idle to base load, the vibrations of the MBK10 shaft of the intermediate block follow immediately the load of the machine. At that very moment, when base load is reached, the maximum vibrations are there. In comparison, the turbine vibrations at bearing MBD10 show a normal temperature-related increase.

In Fig. 3.108, we see a thermal vibration growth at the turbine, and that the vibrations rise with the torque at the intermediate block.

Vibrations rise continuously until turbine trip (see Fig. 3.109). Then they disappear instantly because the torque disappears as well. Because of this torque dependency, we suspected a division error in the SSS clutch.

The SSS clutch had been sent to England for repair. After repairing, this phenomenon disappeared.

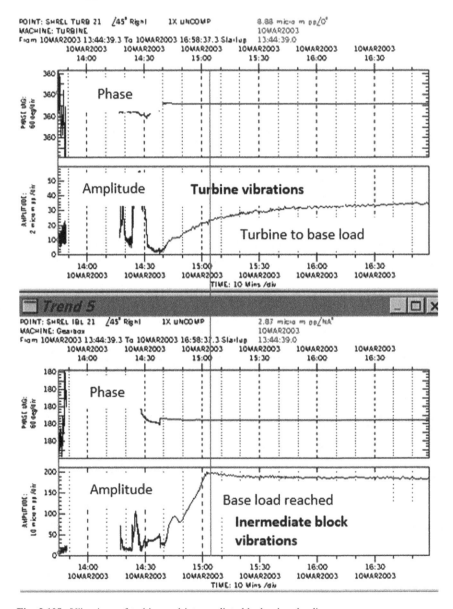

Fig. 3.108 Vibrations of turbine and intermediate block when loading

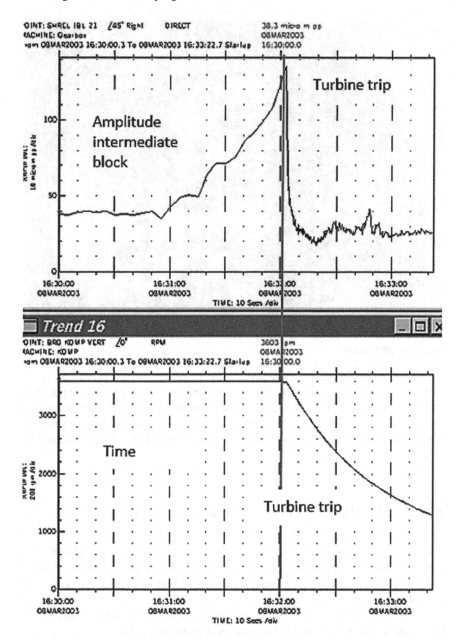

Fig. 3.109 Intermediate block vibrations and turbine speed at trip

Fig. 3.110 SSS clutch

The SSS clutch (see Fig. 3.110) is a **S**elf-**S**peed **S**ynchronizing clutch. It is needed to operate the generator as a motor (without gas turbine) for reactive load (*MVar*) compensation. As soon as there is no torque transmitted from the gas turbine, the clutch will open, and the gas turbine is shut down. The generator runs as a motor.

3.10.2   750 MW Generator in Former Yugoslavia

The driven side bearing shows approx. 40 $\mu m_{s.p.}$ run-out error manifested by a speed-independent amplitude and phase angel at shaft measurement. Run-out vibrations are coming from a geometrical offset and are not governed by the centrifugal force.

Figure 3.111 displays the typical appearance of a run-out error (graph on the left) compared to normal shaft vibrations (graph on the right):

Almost speed-independent vibrations occur because of run-out errors (parallel misalignment, geometrical fault, not driven by centrifugal force). The vibrations are not driven by rotor-dynamic behavior, but the mechanical offset of the error controls the vibrations.

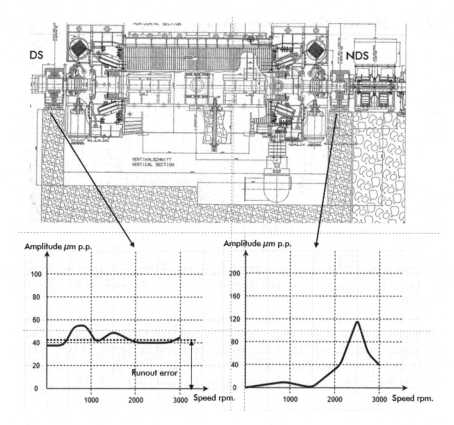

Fig. 3.111 Run-out error at the DS bearing

Figure 3.112 shows the run-out fault of the DS bearing that had been corrected by opening the coupling and correcting its alignment.

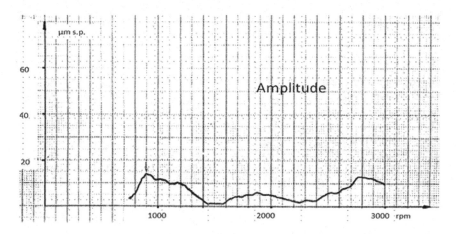

Fig. 3.112 DS bearing vibrations after correction

3.10.3 150 MW Generator in Spain

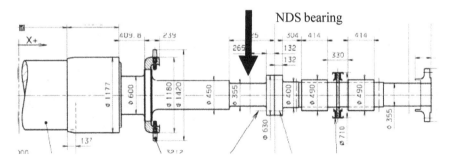

Fig. 3.113 Non-driven side (NDS) shaft of a generator

The shaft at the NDS bearing of a 150 MW generator in Spain (marked by arrow in see Fig. 3.113) became bent and showed a high run-out error in the range of 175 μm.

The following figures show the comparison to the correct situation at commissioning one year earlier. The Bode plot as well as the polar plot shows an offset with the amount of the run-out. The vibrations are dominated by the run-out (Fig. 3.115) unlike the situation earlier without run-out fault (Fig. 3.114) (the run-up vibrations after overhaul do not indicate a run-out error, according to Fig. 3.114).

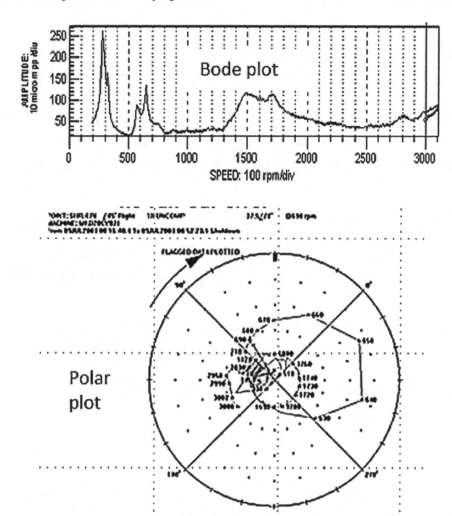

Fig. 3.114 Run-up NDS bearing after overhaul

Figure 3.114 shows a perfect run-out. The polar plot loops, indicating the critical speeds, have their origin close to the center of the graph. The vibration amplitudes also decrease to almost zero between critical speeds.

One year later, we can see a heavy run-out error, manifested by the offset in the Bode and Polar plot (see Fig. 3.115). The bad run-out had been due to a shaft crack, which will be described later in Sect. 3.13.4.

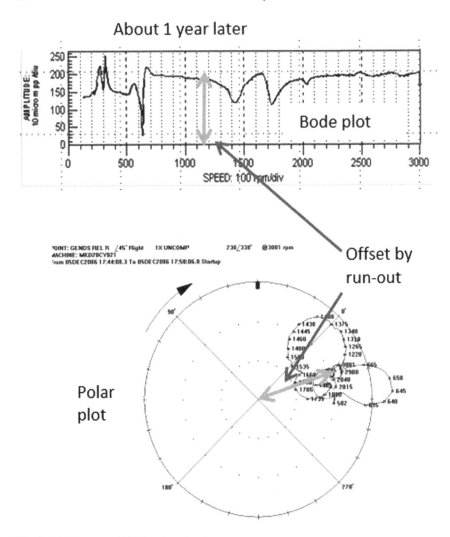

Fig. 3.115 Run-up at NDS bearing after 1 year

We see a considerable run-out error (compare with fair run-out in Fig. 3.114). The polar plot loops, standing for the critical speeds, show an offset of the center of the graph. The amount of the offset indicates the amount of the run-out error.

3.11 Rubbing

3.11.1 80 MW Steam Turbine

Figure 3.116 shows the design drawing of an 80 MW steam turbine.

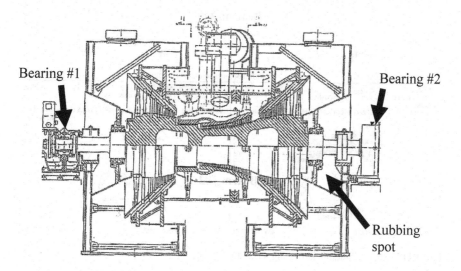

Fig. 3.116 Cross section of steam turbine

Figure 3.117 shows the polar plots of the two bearings; we see rubbing pronounced at the left side bearing #2. At Fig. 3.118, we have an enlarged image of bearing #2 left. This polar plot covers a time span of almost 16 h.

Due to the erratic in-stationary behavior of vibration displacement and phase, balancing is impossible. The rubbing spot must be eliminated, in this case by moving the glands regarding the shaft.

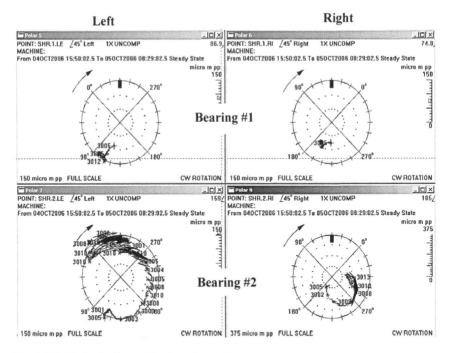

Fig. 3.117 Shaft vibrations at bearing #1 and #2

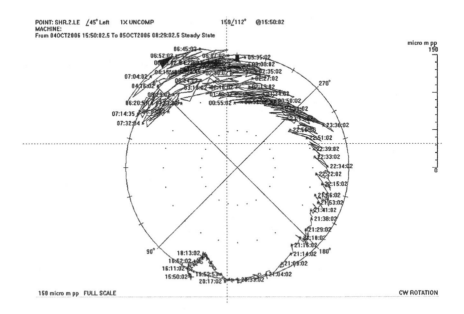

Fig. 3.118 Bearing #2 left enlarged

3.11.2 150 MW Steam Turbine

The following example demonstrates a rub at a 150 MW steam turbine, by means of a Bode plot. It started at high speeds (>3100 rpm). The rub occurred close to bearing #2. Figure 3.119 shows the sudden rubbing occurred close to rated speed. Figure 3.120 is a more detailed view of the rub.

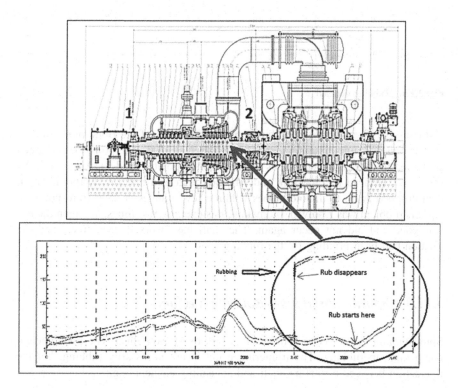

Fig. 3.119 Turbine drawing and Bode plot

Figure 3.119 shows the run-up and run-down vibrations of bearing #2. A rubbing action takes place, indicated by the circle. After reduction of the speed, at 2500 rpm, the rubbing disappears followed by a normal run-down.

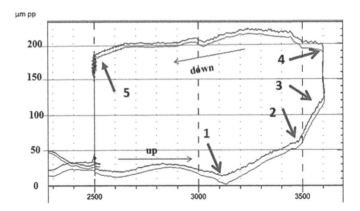

Fig. 3.120 Rubbing detailed

Figure 3.120 explains the rubbing process by means of the Bode plot:

The unit was running up fine, until a rub started at 3100 rpm (point 1). The speed continued to 3600 rpm; vibrations became worse (points 2 and 3). At point 4, the speed was reduced to 2500 rpm and vibration increased first slowly (point 4) but then dropped again to the old values within a few minutes at 2500 rpm (point 5).

This problem was resolved, by making this exercise two times, and then the rub was gone and did not occur again. The shaft has "worked itself free," and the rotation had been grinding off the rubbing spot.

3.11.3 Spiral Vibrations—Rotating Vectors

Spiral vibrations (or vector rotation, Newkirk or Morton effect) is a phenomenon comparable to rubbing. The triggering event is not necessarily rubbing, but the interaction between a rotating shaft and a standing element, e.g.:

- bearing shell,
- seal ring or probably
- slip-ring brush.

At an elastic rotor, due to the elastic response of the shaft to an unbalance, we always will experience an un-symmetric gap, and in case of a bearing, between shaft and bearing sleeve.

At the displacement high point (at the smallest region of the gap), a temperature hot spot develops, which is fixed to the rotating shaft. The consequence of this hot spot is a slight bending of the rotor with its high point at the hot spot. Under certain conditions this hot spot (or bending) becomes in-stationary and starts to travel slowly regarding on the rotating shaft. We will measure an "up and down" of vibrations, in a polar representation: a "rotating vector."

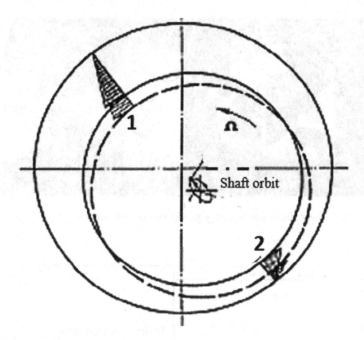

Fig. 3.121 Development of the hot spot

Figure 3.121 indicates how the hot spot develops due to the elastic deformation of the rotating shaft. This hot spot had been measured for the first time in 1977 in Finspong at a large steam turbine company.

The shaft had been equipped with 12 thermocouples approx. 1 mm below the shaft surface, and the signals had been abducted to the standing instrumentation by means of a special slip-ring device.

After having experienced a heavy rotating vector problem at one of the nuclear plants (most likely NDE bearing of the generator), a rotor in an original bearing (450 mm) had been investigated. Figure 3.122 shows the bearing which was used for the tests.

It was found that the normally operating 3-segment tilting pad bearing develops vibrations with 41 $\mu m_{s.p.}$ and diametrical temperature differences of almost 10 K at 3000 rpm (see Fig. 3.123). These temperature differences are enough to distort the journal.

Fig. 3.122 3-segment tilting pad bearing with 450 mm journal diameter

68.14	72.48	66.9
61.49	67.05	64.64
59.79	62.57	60.33

Journal temperature distribution:

Fig. 3.123 Journal temperatures (measures in °C)

We also saw a lagging of the temperature vector behind the vibration vector of approx. 20 angular degrees. This phase lag is responsible for the slowly rotating vibration vector. Figure 3.124 indicates the phase difference between vibration and hot spot.

Fig. 3.124 Temperature and vibration vector of journal

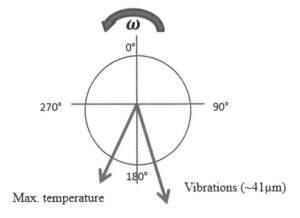

Max. temperature

Vibrations (~41μm)

Here, we experience the second important condition for spiral vibrations:

The phase angle difference between the high point of temperature and the high point of the resulting vibrations is the reason for the vector rotation. If there would be no phase angle difference, this vector would be stationary and could be a subject to compromise balancing (which requires constant phase angels).

As already described, in this case it was found that the spot of max. temperature on the journal will differ from the vibration vector by 20°. To eliminate the spiral vibrations, oil spray nozzles had been introduced between the segments in order to equalize the circumference shaft temperatures (see Fig. 3.125).

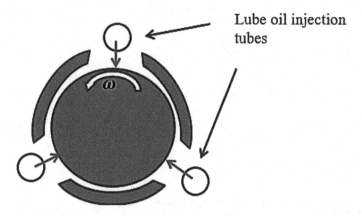

Lube oil injection tubes

Fig. 3.125 Lube oil injection tubes between the segments

The vibration vector can behave like an opening spiral which is very dangerous and can lead to damage. Figure 3.126 shows the opening spiral vibration. The machine tripped at the prepared trip value of 240 $\mu m_{p.p.}$

In most of the cases, we have a constant vector length, which means the radius of a circle in the polar plot. Figure 3.127 shows vector rotation with constant vibration vector.

The vibration behavior is dependent on various parameters, like critical speed, operating speed, bearing type, rotor design, etc.

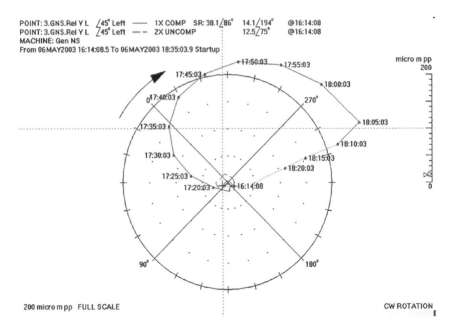

Fig. 3.126 Vector rotation, opening spiral

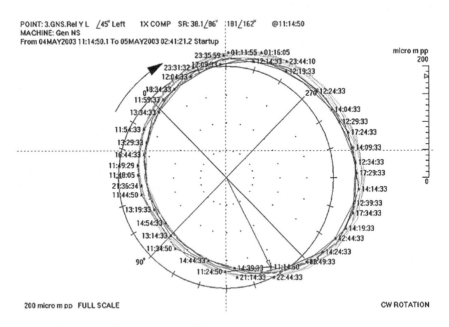

Fig. 3.127 Vector rotation, constant rotating vector

There are cases of the dangerous "opening spiral," (Fig. 3.126) and also of a continuous vector rotation with an unchanged amplitude of the rotating vector Fig. 3.127.

Model of conception out of experience in several plants:

Condition 1:

The shaft line of a turbo-set has several points of contact with stationary structures (bearing journals, shaft seals, slip-ring brushes, gas seals on the shaft, etc.). At all these points, slight rubbing and heat loss are taking place. The synchronous interaction of rotation and shaft vibration produce a situation in which it is always the same point on the shaft which has the minimum distance from the stationary part. This results in a more or less pronounced, unsymmetrical warming of the circumference of the shaft (hot spot) and this, in turn, leads to distortion (bending) and hence vibration.

Condition 2:

One critical speed of the shaft line must lie near the operating speed; it is in almost all cases found just below the operating speed (15%). From the basic principles, we know that as the shaft accelerates through a critical speed, the eccentricity of the shaft's center of gravity changes phase (from 0° to 180° at SDOF shaft) relative to the shaft deflection.

Condition 2 thus results a certain lagging phase difference between the eccentricity of the center of gravity and the shaft deflection at operating speed. Therefore, the high point due to warming (given by Condition 1) continuously lags behind the shaft bending (regarding the phase angle conventions) and therefore causes a continual changing rubbing spot.

A counter-rotating bending and therefore a rotating vector appear superimposed on the "steady" vector of the residual unbalance. The non-stationary rubbing spot (rotor bending) is a consequence of the critical speed close to the operating speed. If there would be no critical speed close to the operating speed, the rubbing spot would be stationary as well as the resulting vibration vector.

Figure 3.128 indicates the kinetic shaft deformation at the slip-ring shaft critical closely below the operating speed at large generators.

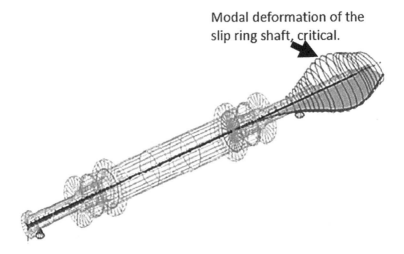

Modal deformation of the
slip ring shaft, critical.

Fig. 3.128 Modal deformation of slip-ring critical

The "spiral vibration" phenomenon is composed of a rotor fixed, but stationary vector. On top of this vector, an instable rotating vector is sitting, caused by a thermal bend slowly rotating in relation to the rotor at rated speed (normally opposite to the rotation sense, more accurately expressed: opposite to the orbit rotation).

There are a few cases when the orbit rotates opposite to the rotation, and the rotating vector goes with the rotation.

Figure 3.129 shows that vector rotation is composed of two vectors.

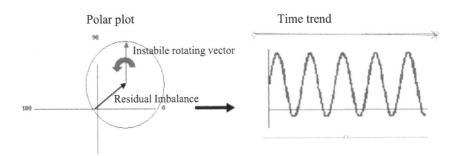

Fig. 3.129 Vector rotation, time trend

3.11.4 750 MW Steam Turbine Generator

In the past, we had spiral vibrations during commissioning at almost every large 2-pole 50 Hz generator. This example shows an especially grave example of a 750 MW steam turbine in Serbia. Figure 3.130 shows the assembly drawing of the generator, and Fig. 3.131 shows a similar generator rotor in the workshop.

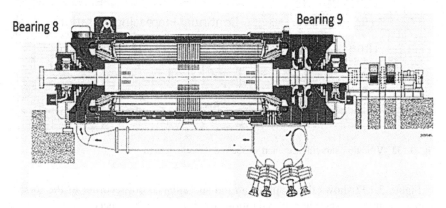

Fig. 3.130 2-pole (3000 rpm) generator

Fig. 3.131 750 MW generator rotor

The vibration surveillance chart of this generator is visible in Fig. 3.132. Looking at the chart, the generator end bearings showed that what we call "an opening spiral," which was indicated in the time trend as a continual increasing wave of vibrations until the unit had to be shut down. Before the vibrations went too high, we had to stop the machine.

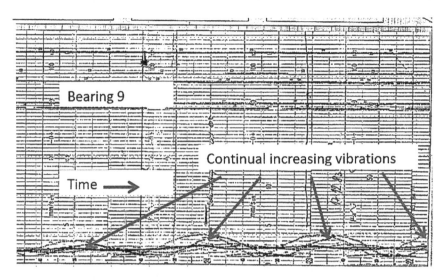

Fig. 3.132 Vibration surveillance chart of the old Amrein system

Figure 3.132 shows the graph of an old surveillance dot recorder of the former Amrein company, showing the vibrations over time at rated speed.

Looking at the chart, the generator end bearings showed that what we call (referring to the polar representation) "an opening spiral," which is indicated in the time trend of Fig. 3.132 as a continual increasing wave of vibrations until the unit had to be shut down. Before the vibrations went too high, we had to stop the machine.

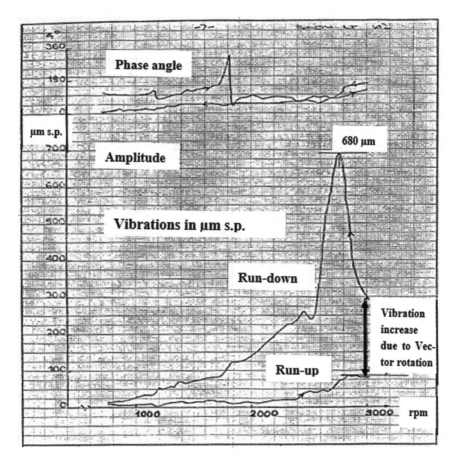

Fig. 3.133 Run-up and subsequent run-down at NDS bearing #9

In Fig. 3.133, we see how high the vector rotation can drive up the run-down vibrations to dangerous values. We see a plot of an X/Y recorder for the run-up and the subsequent run-down. The run-up is fair but at rated speed (3000 rpm), the vibrations start to increase until the machine must be shut down. The run-down reached dangerously high vibrations of 680 $\mu m_{s.p.}$ at the slip-ring bearing. This value corresponds to 1360 $\mu m_{p.p.}$ according to the units of today

In 1982, we had no instruments for visualizing the polar plots, so we had to draw them by hand. The influence of the brushes had been investigated with such a hand-drawn plot (see Fig. 3.134).

The machine was started without brushes showing already a vector rotation against the sense of rotation. Then 33% of the brushes had been attached, and after having attached 66% of the brushes, we ended up in an opening spiral. But since we found vector rotation already without brushes and the brushes are needed for operation, we concentrated on the bearing hot spot and the slip-ring critical. This vector rotation example demonstrates the influence of the slip-ring brushes.

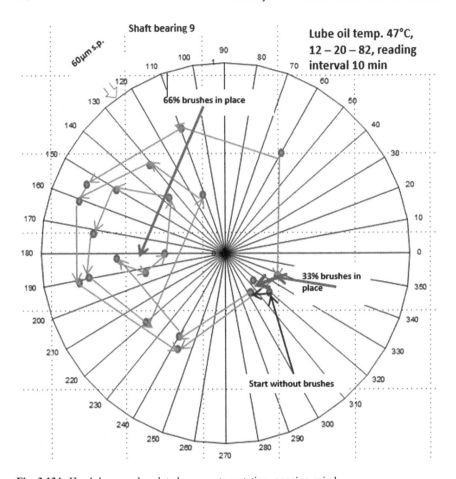

Fig. 3.134 Hand-drawn polar plot shows vector rotation, opening spiral

One of the countermeasures was a movement of the bearing shell of the auxiliary bearing approx. 300 mm away from the generator NDS bearing. This increased the active length of the slip-ring shaft and decreased its critical to increase the distance to the operating speed. Several generators had been modified according to Fig. 3.135.

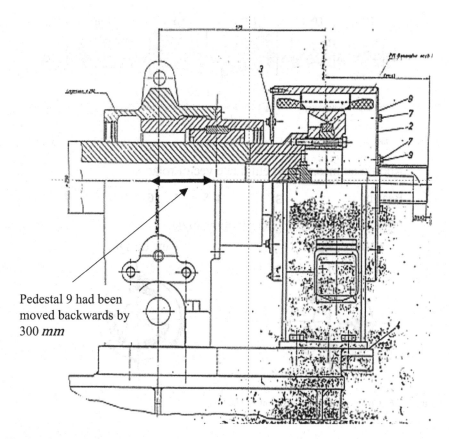

Pedestal 9 had been
moved backwards by
300 *mm*

Fig. 3.135   Modification of pedestal 9

The modification according to Fig. 3.136 had also been realized at various
generators. A photograph of the original 3-segment tilting pad bearing is shown in
Fig. 3.122.

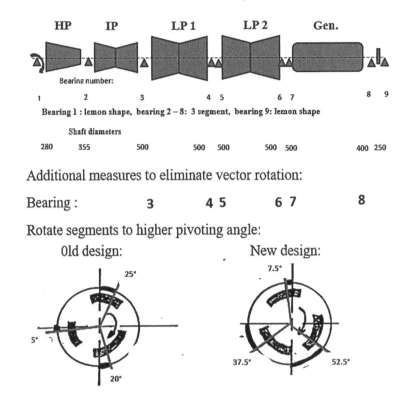

Additional measures to eliminate vector rotation:

Bearing :   3   4 5   6 7   8

Rotate segments to higher pivoting angle:

Fig. 3.136 Change of bearing segment orientation

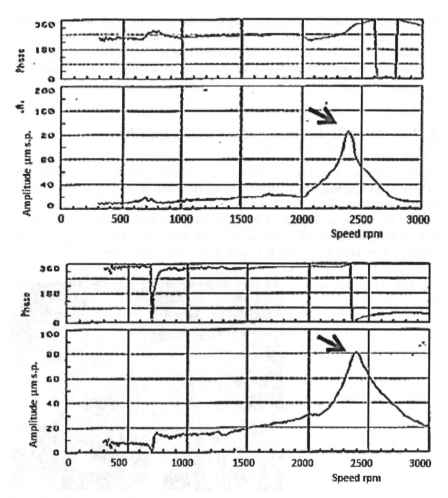

Fig. 3.137 Slip-ring critical vertically and horizontally

In Fig. 3.137, it can be seen that the slip-ring critical had been moved down to 2400 rpm by means of the measures described earlier. Practice showed:

1. Since it was impossible to design an overcritical rotor (regarding the slip-ring critical), that a margin of at least 600 rpm below the operating speed of 3000 rpm was needed.
2. It was important that the margin of 600 rpm was kept vertically and horizontally.

We assumed that the isotropic bearings had been achieved by the segment modification and the lowering of the critical speed mainly by the movement of pedestal 9.

There were about 10 steam power plants between 500 and 1400 MW per turbo-set at which the spiral vibrations had been eliminated successfully:

- by moving the auxiliary bearing away from the generator NDS bearing by about 300 mm to lower the slip-ring critical away from the rated speed and
- modification of the generator NDS bearing to obtain an (almost) symmetric bearing with two carrying segments and achieving a reduced hot spot and more equalized temperatures at the rotating journal.

3.11.5   150 MW Generator Rotor in the Spin Pit

Figure 3.138 shows the rotor of an air-cooled 150 MW generator rotor in a spin pit in Italy. The client required to perform the acceptance tests with the original bearing pedestals from the plant. The rotor was "naked" having no brush-gear, baffles, etc. but still we experienced spiral vibrations-rotating vectors at the slip-ring end. Since there was no other spot of interaction between shaft and standing parts except the bearings, they must have been the cause of the spiral vibrations.

Fig. 3.138 Rotor in the spin pit

The only possibility for the hot spot origin will be the interaction of the shaft with the bearing shell. It must have been in this NDS bearing (see Figs. 3.139 and 3.140).

Fig. 3.139 Non-driven side pedestal

Fig. 3.140 Assembly of NDS bearing

Figure 3.141 shows the NDS bearing sleeve with the jacking oil inlet.

During tests, we discovered the influence of jacking oil on the vector rotation problem, engaging and disengaging the jacking oil during operation at rated speed. Figure 3.142 shows how the jacking oil influences the vector rotation.

Fig. 3.141 NDS bearing sleeve

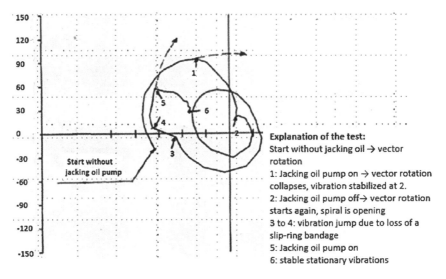

Fig. 3.142 Demonstrates the jacking oil test

Since there was no other "rubbing" element than the interaction between shaft and shell of the NDS bearing, we tried to investigate the influence of the jacking oil at rated speed. The idea was that engagement of jacking oil at speed will provide more lube oil into the bearing which reduces the suspected hot spot. We wanted to prove this bending by means or run-out measurement with dial indicators (see Fig. 3.143).

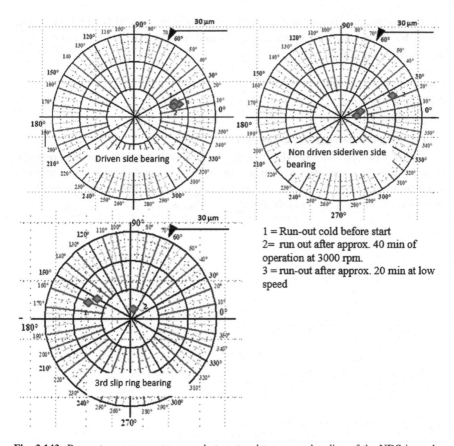

Fig. 3.143 Run-out measurements: prove hot spot and temporary bending of the NDS journal

In the cold condition before start, we measured about:

DS bearing: 12–13 µm
NDS bearing: 7–8 µm
3rd bearing: 15 µm.

Then the rotor had been accelerated to 3000 rpm and kept there for 40 min. After that, it was brought down to low speed and run-out measurement repeated in warmer condition:

DS bearing: 14 µm
NDS bearing: 22 µm
3rd bearing: 2 µm.

DS bearing practically unchanged similar high differences at NDS and 3rd bearing. Then the rotor had been operated for 20 min at 500 rpm and the run-out was measured again cold:

DS bearing: 12–13 μm
NDS bearing: 8 μm
3rd bearing: 12–13 μm.

The values were back at the original cold values.

This test proved that the NDS bearing develops a hot spot at normal operation which results in a considerable bending of the slip-ring end. At first, we wanted to equalize the hot spot by increasing the carrying sector of the bearing sleeve from 60° to 120°, but this had no effect at all (see Fig. 3.144).

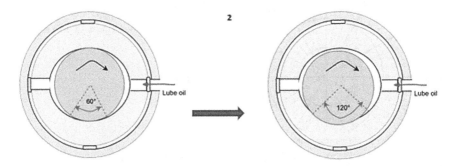

Fig. 3.144 Bearing modification 1st attempt

After having opened the 2nd lube oil inlet at the outer and inner bearing ring (see Fig. 3.145), the journal received oil now from both sides. The rotating vector effect disappeared completely, as can be seen in the final measurements shown in Fig. 3.146.

According to Fig. 3.146, we measured a stable, stationary time trend and practically no difference between a cold run-up and a hot run-down at a much lower vibration level.

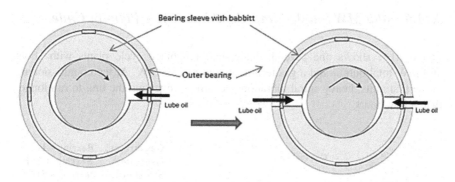

Fig. 3.145 Successful modification

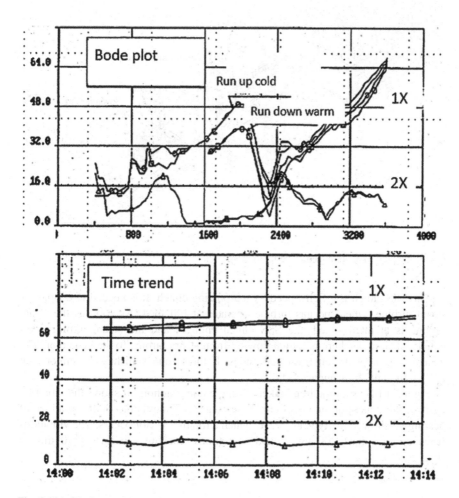

Fig. 3.146 Final measurements

3.11.6 400 MW Single-Shaft Combined Cycle Plant in Chile

Figure 3.147 shows one several single-shaft combined cycle plants with about 400 MW total output and a gas turbine output of 250 MW. At the commissioning of the first units, heavy spiral vibrations occurred preventing the unit to run longer than a few hours.

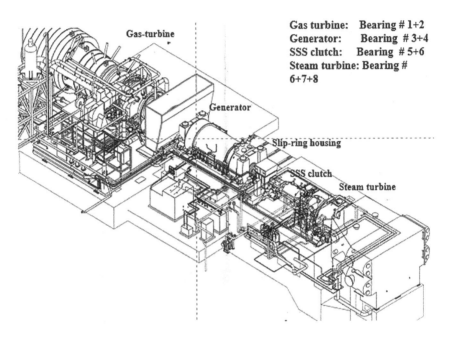

Gas turbine: Bearing # 1+2
Generator: Bearing # 3+4
SSS clutch: Bearing # 5+6
Steam turbine: Bearing #
6+7+8

Fig. 3.147 400 MW combined cycle turbo-set

The SSS clutch is a **S**elf-**S**peed **S**ynchronizing clutch. It is needed to couple the steam turbine, when the gas turbine operates at full load. As soon as the steam turbine is at rated speed, the SSS clutch engages and the steam turbine stays coupled as long as there is torque flowing from steam turbine to the generator. The generator is now driven from both sides. If the torque vanishes (generator shut down), the clutch disengages.

Figure 3.147 is a time trend taken during commissioning. It shows the vibration trend at the generator side SSS clutch bearing #5. The relative shaft vibrations of bearing #5 showed the typical "wave" when a rotating vector occurs. The vibrations reach up to almost 200 $\mu m_{p.p.}$ and sometimes reaching trip level (240 $\mu m_{p.p.}$) (Fig. 3.148).

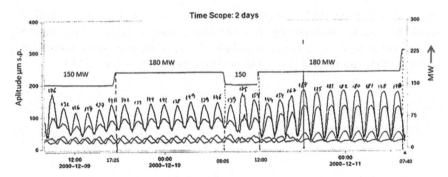

Fig. 3.148 Vibrations at generator side SSS clutch bearing (peaks show max. values of rotating vector activity of Fig. 3.149)

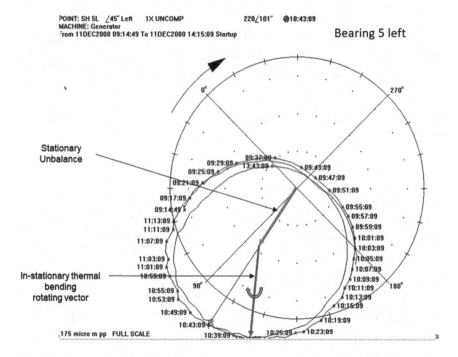

Fig. 3.149 Polar plot with offset from center

Figure 3.149 demonstrates the vector rotation circle, with its offset from the graph center by the amount of unbalance.

According to our former experiences, we suspected a critical speed close or slightly below the operating speed. We produced a block trip because we needed a run-down in the coupled condition to measure the relevant critical speeds. At a normal trip, the SSS clutch would disengage, and the system would have changed.

There was a pronounced critical speed in the vertical direction exactly (or slightly above) at 3000 rpm and in the horizontal direction at 2800 rpm. According to the information from the design people, this was the critical of the slip-ring shaft and should be close to 4000 rpm. Figure 3.149 demonstrates a critical speed of the SSS clutch shaft at rated speed vertically and closely below the rated speed horizontally, *which is visible in the bode plot close to the operating speed* (Fig. 3.150).

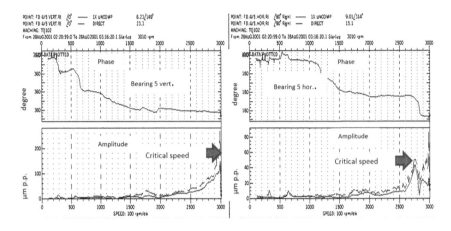

Fig. 3.150 Run-down at block trip engaged SSS clutch (critical speed at and close to operating speed)

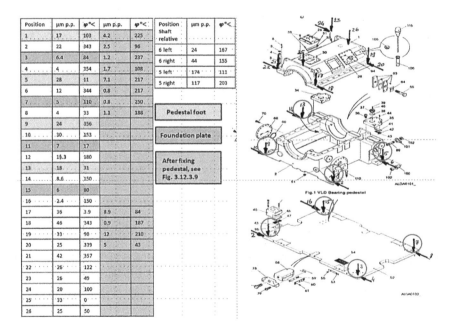

Position	μm p.p.	𝜑°<	μm p.p.	𝜑°<	Position	μm p.p.	𝜑°<
1	17	103	4.2	225	Shaft relative		
2	22	343	2.5	96	6 left	24	167
3	6.4	84	1.2	237	6 right	44	155
4	.4	354	1.7	108	5 left	174	111
5	28	11	7.1	217	5 right	117	203
6	12	344	0.8	217			
7	5	110	0.8	250			
8	4	33	1.1	188	**Pedestal foot**		
9	24	356					
10	30	153			**Foundation plate**		
11	7	17					
12	15.3	180			**After fixing pedestal, see Fig. 3.12.3.9**		
13	18	31					
14	8.6	150					
15	6	80					
16	2.4	150					
17	36	3.9	8.9	84			
18	46	343	0.9	187			
19	33	90	12	210			
20	25	339	5	43			
21	42	357					
22	26	122					
23	26	49					
24	20	100					
25	33	0					
26	25	50					

Fig. 3.151 Vibration mapping of the pedestal structure

The vibration mapping in Fig. 3.151 shows looseness of the SSS clutch pedestal. On the right side, we see a sketch of the SSS clutch pedestal with numbered arrows, the measuring points. The left side indicates the table with the measured values at these points. The yellow and green entries of the table mark the different elements of the pedestal assembly.

The four corner blocks of Fig. 3.152 are foreseen to hold the pedestal down.

Fig. 3.152 SSS clutch pedestal

Figure 3.153 shows one of the 4 corner blocks, which should hold down the pedestal and prevent relative motion of the pedestal against the foundation plate.

At all four corners, we find big differences of vibration between pedestal foot and foundation plate (see points 1 and 3; 5 and 7; 9 and 11; 13 and 15 in Fig. 3.151, marked with red circles). The higher vibrations of the pedestal foot indicate looseness.

We therefore manufactured a clamp see Fig. 3.153, to improve the contact from the pedestal to plate. We therefore manufactured a clamp (see Fig. 3.153), to improve the contact from the pedestal to plate.

Fig. 3.153 Temporary test clamp

The test result proved that the diameter of vector rotation circle became smaller and we were on the right track (see Fig. 3.154).

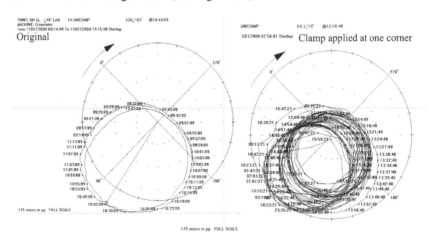

Fig. 3.154 Comparison of polar plots

After the experiment with the clamp, we decided to eliminate the relative motion underneath the relevant bearing #5 by jacking up the pedestal. The adjusting screws of the "elephant feet" (see Fig. 3.155) had been moved by three clicks, which correspond to 0.36 mm upward. The pedestal had been clamped now with good contact to the foundation. The vector rotation was practically eliminated (see Fig. 3.156). The phase angle of the remaining was now completely different, which let us assume that we have changed the critical speed.

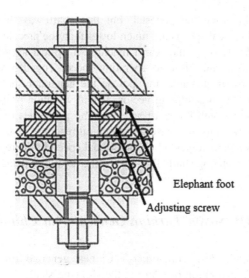

Fig. 3.155 "Elephant foot" design

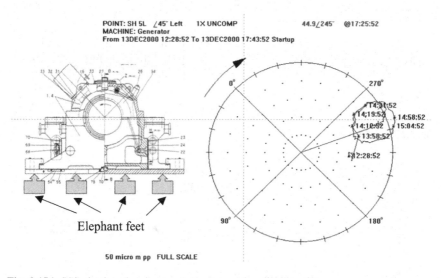

Fig. 3.156 SSS clutch pedestal support structure, polar vibration plot

Our message to the technical department from jobsite:

This measurement shows that the bearing pedestal can practically move kinetically with very little restrict vibration by its foundation, which surely was not the basic idea. Considering this fact, the following scenario becomes very likely:

At the rotor-dynamic calculation, the fact of an almost freely moving pedestal had of course not being taken into consideration. It was assumed that the weight of the shaft will hold down the pedestal, but this shaft was thermally distorted. Therefore, the real critical speeds are much lower than the precalculated ones at the warm machine. We have at the warm machine a critical speed of the shaft in bearings #5 and #6 at and close to operating speed. Together with the unstable unbalance caused by an unstable hot spot on the generator shaft end at the SSS clutch side, we end up in a rotating vector—spiral vibration problem.

Since we must live with the shaft and coupling properties (which are creating this phenomenon), we must decrease the structural response resp. the sensitivity of our system, which means to get rid of the critical speed by increasing the pedestal stiffness seen by the rotating shaft. This will be the most promising countermeasure.

3.11.7 300 MW Steam Turbine Generator in China

Figure 3.157 shows the design drawing of a Chinese generator rotor, and Fig. 3.158 documents the vector rotation problem at the slip-ring shaft.

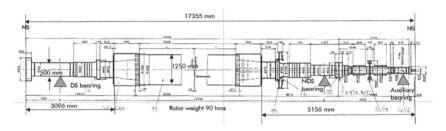

Fig. 3.157 Generator rotor

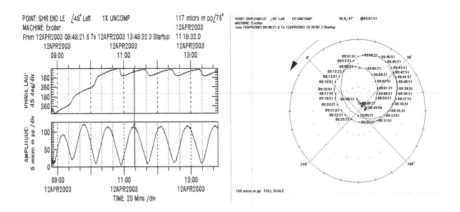

Fig. 3.158 Time trend and polar plot

Improving measures which eliminated the vector rotation:

- exchange of brushes against Union Carbide NKF 634 and
- increase in oil-flow in NDS bearing to over 4 l/s (originally only 2–3 l/s had been measured).

These measures cured the rotating vector problem, as can be seen in Fig. 3.159. The hot spot at the slip-ring shaft had been reduced rather than shifting a critical speed.

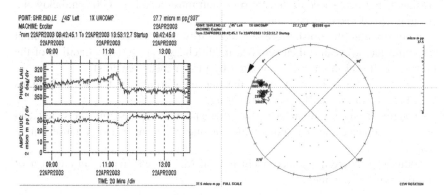

Fig. 3.159   Time trend and polar plot after the correcting measures

3.11.8   General Conclusions of Vector Rotation—Spiral Vibrations

According to our common comprehension now, we interpret vector rotating—spiral vibrations as follows:

A hot spot on the shaft creates a temporary in-stationary bending (an unbalance) due to the high mobility of the shaft this unbalance becomes instable. This instability is triggered by the phase lag of the thermal hot spot behind the vibration high point, due to the fact of a critical speed close below the operating speed.

- As a general countermeasure, the slip-ring critical must move away from the rated speed (down by at least 600 rpm).
- The rotating vectors occur almost exclusively at the slip-ring shafts of generators.
- Rotating vectors seem to be a specialty of a design where the slip-ring shaft is relatively long compared to the active length of the generator rotor and the consequence is a slip-ring shaft critical just below the rated speed. At rotors where this is not the case, those from other design principles very rarely have this problem to this extend.

Slip-ring shafts are far too slender (probably a legacy from design, see Fig. 3.157) and have their critical mostly just below the operating speed:

- For 50 Hz generators (operating speed at 3000 rpm), the critical is around 2800 rpm.
- The 60 Hz generators (operating speed at 3600 rpm) do not have this problem, because the slip-ring critical is too far below the operating speed.

We proposed to change the design in a way that the critical speed of the slip-ring shaft part is above the operating speed at 50 Hz generators. In case this is not possible, the critical speed of the slip-ring shaft must be at least 600 rpm below the operating speed (experience value).

In our opinion, the brushes are not necessarily the root cause, but a substantial worsening factor because their removal might change the system condition of the slip-ring shaft (but this must be checked by further measurements).

The root cause is in many cases the uneven warming of the shaft along its circumference in the NDS generator bearing.

Changing from one oil inlet to two diametrically opposed oil inlets, left and right, will equalize the hot spot of the rotating shaft and vector rotation might disappear.

Additionally, here are some conclusions of Mr. Alfred Ziegler, an engineer of our company in Mannheim:

- Hot spot stability calculations show a significant increase in stability margin by introduction of 5 pad tilting pad bearings.
- The change of stability margin is mainly influenced by increase in slip-ring shaft horizontal critical speed.
- Introduction of 5 pad tilting pad bearings on both ends of the slip-ring shaft reduced the diameter of the spiral from more than 400 to 41 $\mu m_{p.p.}$.
- Introduction of "low friction brushes" reduced the spiral diameter further 20–30%.

3.12 Development of 2X (Twice Rotation Frequency) Vibrations

3.12.1 Sag Excitation

2X (twice rotation frequency) vibration generally occurs only in 2-pole generators. Because of the slot distribution, the generator has a polar moment of inertia in the pole zone which is different to that in the winding zone (see Fig. 3.160); regarding stiffness, it is thus anisotropic. The pole zone is stiffer than the winding zone.

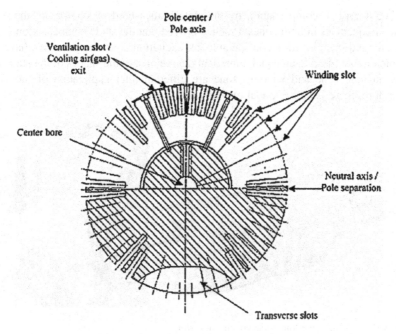

Fig. 3.160 Cross section of a 2-pole generator rotor

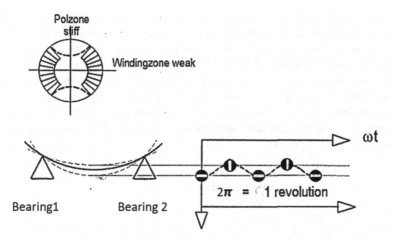

Fig. 3.161 Describes sag excitation

The sag line changes according to Fig. 3.161 and produces a 2X vibration. The transverse slots in the pole zone will equalize the sag line difference, but only in a "static" way. That means that same amount of sag will be obtained in the winding and the pole zone.

The degree of compensation by the slots is measured by comparing the resonance frequencies in both zones. Practice proved that the static compensation is not enough, and there is still a considerable 2X excitation at some rotors. We have the suspicion that (despite an equal numerical degree of sag) the shapes of the static sag lines in pole zone and winding zone are different after application of the slots. Therefore, there is still a modal excitation at speed.

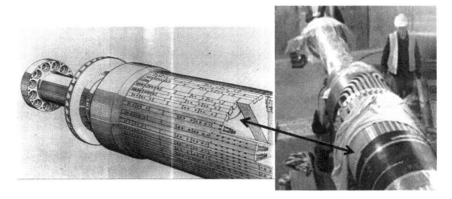

Fig. 3.162 Indicates the transverse equalization slots

400 MW generator NDS generator bearing:

The 1st critical speed will be excited by sag line change at half of the unbalance driven critical speed. The transverse equalization slots shown in Fig. 3.162 distributed along the active length of the rotor should compensate the sag change during rotation, but only in a "static" way (as described above).

Practice proved that the static compensation is not enough, and there is still a considerable 2X excitation at some rotors. We have the suspicion that (despite an equal numerical degree of sag) the shapes of the static sag lines in pole zone and winding zone are different after application of the slots. Therefore, there is still a modal excitation at speed.

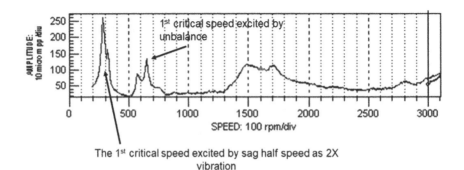

Fig. 3.163 Run-up of a 150 MW generator rotor

In Fig. 3.163 we see that a compensated rotor still develops a considerable 2X vibration. At a similar rotor (see Fig. 3.164), the equalization slots had been machined deeper and the rotor had been tested in the spin pit.

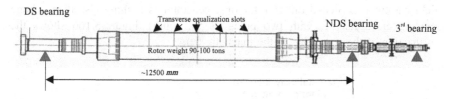

Fig. 3.164 Indicates a large 2-pole generator rotor

The test in the spin pit (see Fig. 3.165) should document the effect of the transverse equalization slots. The graphs show the filtered 2X relative shaft vibrations at the bearings DS, NDS and 3rd bearing in the vertical direction. We see that deeper slots reduce the 2X vibrations. The reduction of 2X will only continue, until the phase changes. From then on, the 2X vibration will increase again.

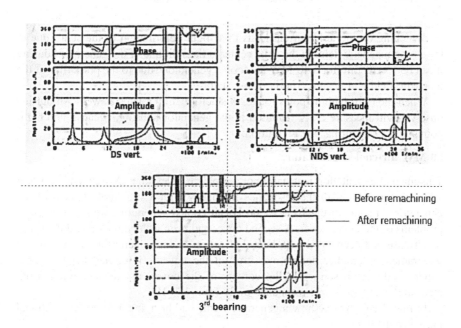

Fig. 3.165 Spin pit test effect of deepening the transverse slots

3.12.2 *Magnetic 2X Vibrations*

Another source of 2X vibrations are coming from the magnetic field of the generator rotor of a 2-pole machine, but this 2X vibration only appears, as soon as the field current is switched on. A rotor of a generator is nothing else than a huge rotating electromagnet with two poles (north, south). Figure 3.166 shows the ovalization of the stator due to the magnetic field.

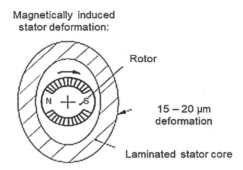

Fig. 3.166 Indicates the stator deformation by the magnetic field

Switching on the field current leads to a slight deformation of the stator due to the strong magnetic field. In the axial center at the split line of the generator housing it can be 15–20 μm (for a 40 MW air cooled generator), due to the properties of the magnetic field and design criteria of the stator.

When the rotor with its poles rotate, this deformation rotates as well. An outside observer will measure a 2X vibration at the generator housing, because two high-points will pass during one revolution.

80 MW air-cooled generator:

This magnetic excitation is usually not a problem unless the 2X frequency (100 or 120 Hz) coincides with the resonance frequency with a structural element of the generator. At some of the 70 MW, 60 Hz air-cooled generators, we had resonance problems of the upper cover. A second cover had been welded on top and the space in between had been filled with buckshot pellets or sand, increasing the weight on the frame. As another countermeasure, all the cover screws had been opened, leaving only the 8 screws in the corners of the cover in fixed condition. Both measures tuned the frame resonance away from 120 Hz.

In Fig. 3.167, we see how a generator frame had been detuned by an additional mass (sand filling).

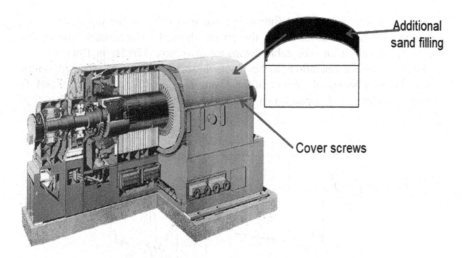

Fig. 3.167 Detuning measure of the frame

250 MW hydrogen-cooled generator:

This is a case where a generator was destroyed by magnetic vibrations. It was a hydrogen-cooled 200 MW generator for a Canadian power plant. It was a special design, where the hydrogen coolers had been placed horizontally on top of the stator core within the housing. This led to an "eccentric" design of the core against its housing. Figure 3.168 shows the cross-section drawing of the generator.

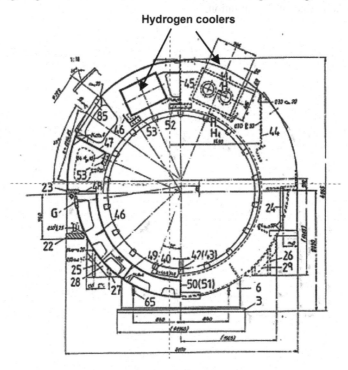

Fig. 3.168 Frame of 250 MW generator, cross section

The acceptance impact test (impacting and measuring at the top center of the housing), which was required by the client, showed an acceptable mechanical flexibility as well as an acceptable resonance at approx. 170 Hz in 1983.

After 3 years of operation in the plant, these quantities had changed significantly. The mechanical flexibility of the casing increased by a factor of 6 and the resonance frequency dropped down to 130–140 Hz (see Fig. 3.169).

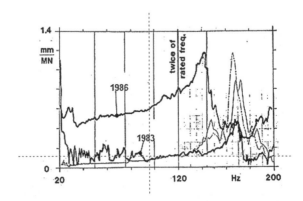

Fig. 3.169 Impact tests of generator frame

In Fig. 3.170, we see the four deformation phases of the frame in operation during one period of 2X vibration.

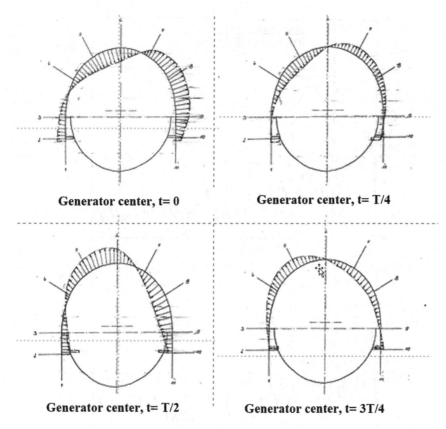

Generator center, t= 0   Generator center, t= T/4

Generator center, t= T/2   Generator center, t= 3T/4

Fig. 3.170 Generator frame deformation

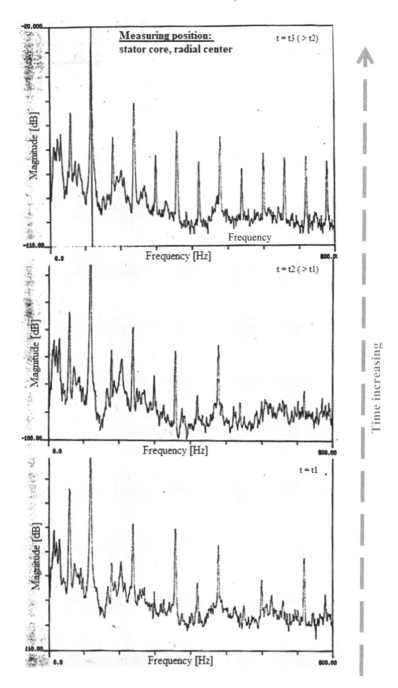

Fig. 3.171 Change of spectrum by time

The spectra in Fig. 3.171 indicate how the frame became loose by time, manifested by the increasing harmonics in the spectrum.

The screws inside the generator frame, which hold the outer casing together with the frame of the laminated iron core, are getting more and more loose because of the 2X vibrations.

Since the screwed connection between core and housing inside the machine was not accessible, the only way to control the increasing housing vibrations was by extending the outside stiffening ribs of the frame. Figure 3.172 shows the first set of additionally welded ribs.

Fig. 3.172 Generator frame, stiffening ribs, 1st attempt

This was done two times, until the height of the ribs was 500 mm. But "calming" the housing lead to higher stress and relative motion of the screwed connection to the core.

Fig. 3.173 Generator frame, stiffening ribs, 2nd attempt

At the next approach (see Fig. 3.173), the ribs had been extended to 500 mm.

Inside the generator housing, we found cracks and cracked material parts. Through the manholes, we could see sparking between core and housing when the machine was in operation. The connecting elements between stator core and casing, which were not accessible after assembly, became looser by time. The final consequence was to build a new generator.

3.13   Rotor Cracks

3.13.1   *Considerations Regarding Lateral Cracks*

Following a few fundamental considerations regarding a lateral rotor crack:

- A lateral crack causes 1X (rotational) as well as 2X (twice rotational) vibrations as main frequency components (the literature also mentions higher harmonics than 2X but in our practice experience they were insignificant).
- In all the cases we experienced, one of the features of lateral cracks is the sudden appearance of a 2X component. The mechanism is very similar to the sag excitation described in Sect. 3.12.1 sag excitation.
- A rotor with a uniform cross-section polar stiffness (like turbines or also a 4-pole generator) will start to produce 2X vibrations in case a crack occurs, because then the polar stiffness is not uniform anymore and the sag line will change two times during one revolution.
- The 2X vibrations are independent of the unbalance situation.
- The 1X vibrations are dependent on the relationship of the circumference position of the crack to the residual unbalance. The vibrations by the crack propagation cause an increasing vibration vector on top of the stationary residual imbalance vector. This overlapping can lead to an immediate vibration increase or a decrease in the first stage of the crack regarding the resultant vector, which we measure.

Siemens made a test with a LP rotor, which was cracked close to its center (see Fig. 3.174). Shaft vibrations had been measured at the left and right bearing and in the center (X and Y are standing for two 90° transducers in one measuring plane). An evenly distributed unbalance had been placed at the same side and at the opposite side of the crack.

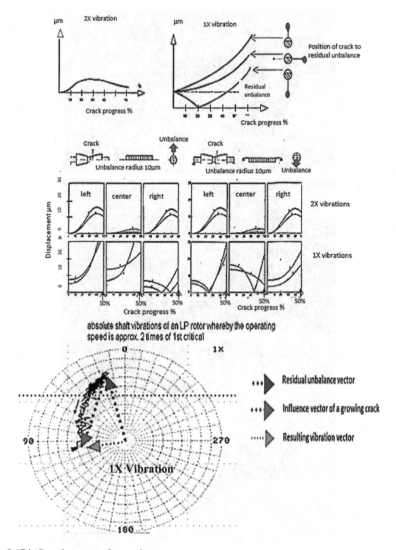

Fig. 3.174 Development of a crack

The polar plot in Fig. 3.174 shows that it is possible that a growing crack may also result in decreasing 1X vibrations at its beginning, depending on the phase relation between unbalance and crack induced vibrations. In Fig. 3.175, we can see the consequence of an undiscovered rotor crack.

Fig. 3.175 Disintegration of a 500 MW turbo-set

3.13.2 4 × 930 MW Steam Turbine Plant, Lateral Crack

The first crack case study deals with a lateral crack in a generator rotor of unit #2. There are four units in a row, and each one has about 930 MW. Every unit has a high-pressure stage, three low-pressure stages and the generator. The shaft is supported by seven bearings. The length of the shaft train is about 70 m. The reactor is a Canadian CANDU heavy water reactor.

Figure 3.176 shows the machine hall, where all four units are located. This hall is over 1 km long.

Figure 3.177 shows the generator. It is hydrogen cooled and has four poles respectively 1800 rpm at 60 Hz.

Fig. 3.176 The machine hall

Fig. 3.177 Generator of unit #2)

Figure 3.178 shows the generator rotor, which has a weight of about 200 metric tons.

Fig. 3.178 Generator rotor hanging on the crane

Our company at that time did not equip the turbo-sets with shaft vibration measurement; therefore, a vibration expert of the client (Alan Sanderson) developed a relative shaft vibration surveillance system, based on a HP desktop computer (personal computers were not capable enough in 1990).

He installed this system at unit #2 in order to test its functionality. At every bearing, there were two proximity probes (left = X, right = Y) and one vertical reference probe (key phasor) for picking up the once-per-revolution speed signal. This installation is shown in Fig. 3.179.

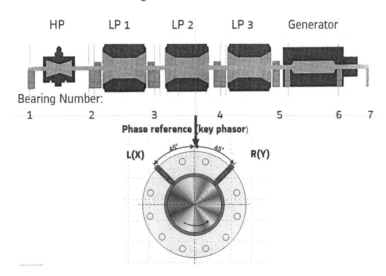

Fig. 3.179 Set up for shaft vibration measurement

The unit was running on base load until a trip happened because of a high-water level in the hot-well (the hot-well is a reservoir in the condenser to collect excess water). There was a short outage of one day planned, to correct the fault. During this time, Alan Sanderson had the opportunity to look at the data his test system had gathered so far. According to the measures (see Table 3.5), the unit was running perfect until approx. 60 h prior to the trip.

Table 3.5 Vibration readings of unit #2 before the trip and before vibration increase

00:45:23		5 Mar 1990			1800 rpm	
POSIT	TOTAL	AMP 1X	ANG 1X	AMP 2X	ANG 2X	GAP
S1L	32.0	20.5	92.0	5.4	223.0	0.0
S1R	32.0	18.7	124.0	2.1	356.0	0.0
S2L	27.0	4.8	27.0	2.2	270.0	0.0
S2R	21.0	6.1	163.0	3.0	56.0	0.0
S3L	17.0	9.2	102.0	4.1	74.0	0.0
S3R	15.0	6.7	232.0	1.1	12.0	0.0
S4L	30.0	17.4	296.0	8.3	41.0	0.0
S4R	22.0	13.3	41.0	4.5	212.0	0.0
S5L	32.0	19.1	302.0	5.6	127.0	0.0
S5R	22.0	14.4	352.0	1.2	29.0	0.0
S6L	28.0	12.8	321.0	6.2	134.0	0.0
S6R	17.0	7.8	39.0	1.3	199.0	0.0
S7L	33.0	22.3	299.0	4.9	67.0	0.0
S7R	23.0	15.4	31.0	1.2	283.0	0.0
ECC	73.0	53.1	108.0	14.3	326.0	0.0
***	50.0	35.6	106.0	10.5	326.0	0.0

All vibration values in $\mu m_{p.p.}$
S1 to 7L = shaft vibration 1 to 7 left
S1 to 7R = shaft vibration 1 to 7 right
ECC = eccentricity, ANG = phase angle, AMP = amplitude, TOTAL = unfiltered

Within the last 60 h before the hot-well trip, there was a massive increase in shaft vibrations at bearing #5 (DS generator). Also, bearing #7 increased (SR bearing) but bearing #6 (NDS generator) stayed constant (see Table 3.6 and Fig. 3.180).

Table 3.6 Vibration readings after the trip and after vibration increase

06:53:23		8 Mar 1990			1799 rpm	
POSIT	TOTAL	AMP 1X	ANG 1X	AMP 2X	ANG 2X	GAP
S1L	22.0	14.8	95.0	1.7	191.0	0.0
S1R	26.0	15.9	124.0	2.1	58.0	0.0
S2L	28.0	9.8	63.0	2.8	255.0	0.0
S2R	18.0	3.4	104.0	2.7	55.0	0.0
S3L	24.0	13.5	176.0	3.9	74.0	0.0
S3R	29.0	17.6	199.0	0.9	350.0	0.0
S4L	37.0	23.6	240.0	8.2	44.0	0.0
S4R	33.0	15.1	267.0	4.8	211.0	0.0
S5L	142.0	132.2	231.0	2.5	90.0	0.0
S5R	102.0	95.7	284.0	1.6	313.0	0.0
S6L	22.0	7.9	216.0	6.6	113.0	0.0
S6R	19.0	15.2	132.0	5.2	238.0	0.0
S7L	86.0	78.1	251.0	5.3	305.0	0.0
S7R	94.0	86.7	353.0	10.6	32.0	0.0
ECC	91.0	65.1	103.0	24.5	23.0	0.0
***	59.0	40.2	111.0	17.8	21.0	0.0

All vibration values in $\mu m_{p.p.}$
S1 to 7L = shaft vibration 1 to 7 left
S1 to 7R = shaft vibration 1 to 7 right
ECC = eccentricity, ANG = phase angle, AMP = amplitude, TOTAL = unfiltered

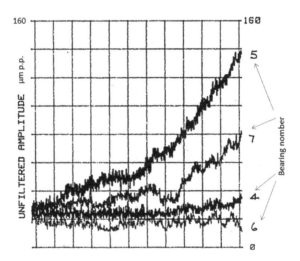

Fig. 3.180 Vibration increase within 60 h

Looking at the run-down vibrations before that trip happened (see Fig. 3.181), we see a fair vibration behavior over the entire speed range here at bearing #5 as well as at bearing #6.

At bearing #6 the shaft vibrations are dominated by the small, normal run-out error (flat response over the entire speed range), which could indicate that the shaft measurement position is very close to a nodal point of the rotor. mode shape.

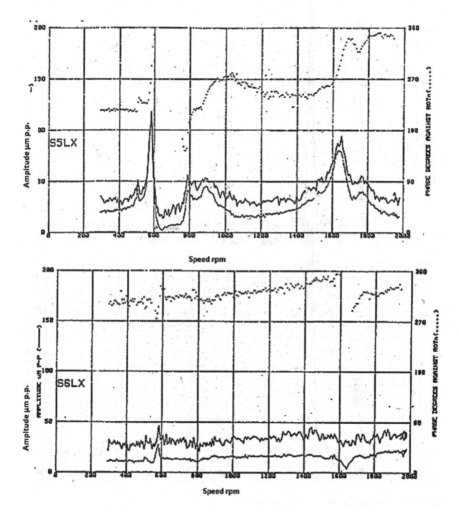

Fig. 3.181 Run-down bearing #5 and #6 before the trip

Figure 3.181 shows a graph of the run-down vibration, taken with the test system of the vibration expert of the plant.

Looking at the run-down vibrations before that trip happened (see Fig. 3.181), we see a fair vibration behavior over the entire speed range at bearing #5 (top) as well as at bearing #6 (bottom) with max. vibrations of about 100 $\mu m_{p.p.}$ at bearing #5. At bearing #6, the shaft vibrations are dominated by the small, normal run-out (flat response over the entire speed range), which could indicate that the shaft measurement position is very close to a nodal point of the rotor mode shape.

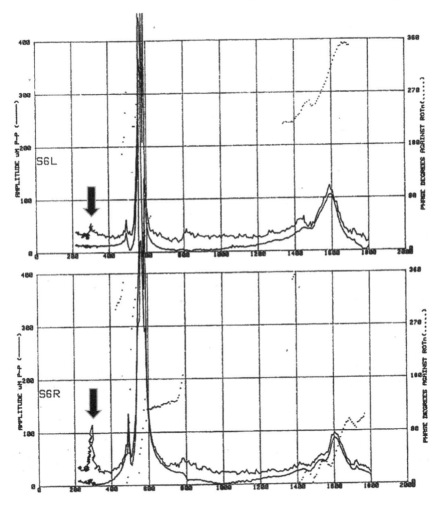

Fig. 3.182 Run-down after trip had occurred

In Fig. 3.182, we see the run-down vibrations after the hot-well trip at bearing #6 (top: left side, bottom: right side). The displacements passing the critical speeds are much higher now, especially at the 1st critical speed at around 550 rpm. At bearing #6, we saw no critical before and now the 1st critical develops vibrations up to the end of the measuring scale (over 800 $\mu m_{p.p.}$).

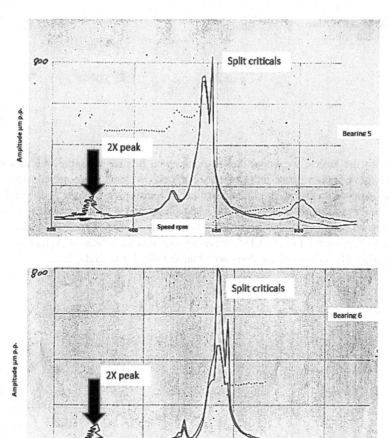

Fig. 3.183 Run-down vibrations at 1st critical 550 rpm

In Fig. 3.183, we see a completely different run-down vibration signature. The vibrations had worsened significantly. It is also conspicuous that we see now a vibration peak at approx. half of the 1st critical speed (black arrows).

This peak occurs only in the overall vibration graph and not in the 1X graph, which means that this peak has a different rotational frequency 1X. The 1st critical speed is "split" into several vibration peaks. It is also conspicuous that we see now a vibration peak at approx. half of the 1st critical speed (arrows). This peak occurs only in the overall vibration graph and not in the 1X graph, which means that this peak has not 1X (rotational) frequency. Since it occurs at ½ of the 1st critical speed, it is very likely that its frequency is 2X (black arrows in Figs. 3.182 and 3.183).

At a 4-pole generator rotor, it is very unusual that there is a 2X vibration. The sudden appearance of the 2X component led us to the suspicion that there is a rotor crack. The vibration levels at bearing # 5 and #6 would suggest that the crack is close to bearing #5. Figure 3.184 shows the region of vibration excitation.

Fig. 3.184 Mode shapes at 1st and 2nd critical

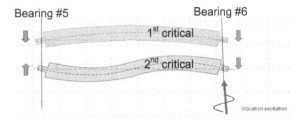

From the basics, we know that response of a flexible structure like at a rotor keeps the vibration phase independent of the mode shape if we are considering a point at or very close to the kinetic excitation region (see Chap. 1.3, Fig. 1.28).

Since the rotor responds with the same phase at 1st and 2nd critical speed at bearing #6 and antiphase at bearing #5, the kinetic excitation force is induced at or close to bearing #6. In Fig. 3.185, it can be seen that the vibration excitation occurs at bearing #6 because phase does not change from 1st to 2nd critical.

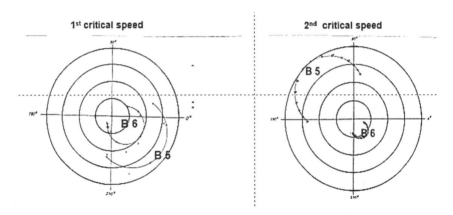

Fig. 3.185 Polar plots of bearing #5 and #6 at the 1st and 2nd critical

Fig. 3.186 Cracked rotor

After opening the manhole at the NDS end of the generator bearing #6, we could see a massive crack at the rotor right away. It started at the hole of one of the conductor cross-bolts, which carry the excitation current to the field winding (see Fig. 3.186). The crack was close to bearing #6 (see Fig. 3.187). Figure 3.188 shows the arrangement of the conductor bolts.

Material scientists had calculated that at the next start the 200-ton rotor would have disintegrated. Four newly designed rotors had been built and exchanged one by one.

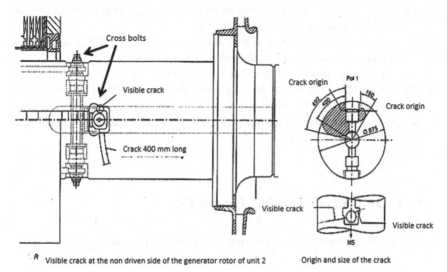

Fig. 3.187 Crack location

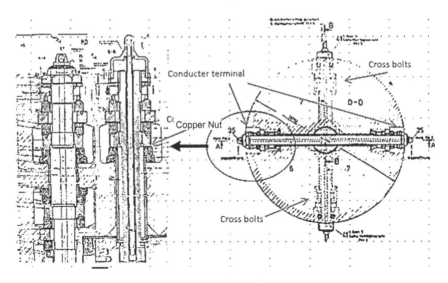

Fig. 3.188 Cross-section drawing of the ND end rotor end

Conclusions:

- Out of the steady state operation, the crack would never had been diagnosed from the vibration signature. Only the transient run-up–run-down behavior gave a clear indication.
- It was necessary to measure shaft vibrations. The bearing cap vibrations would have been too insensitive by far.
- Progressive vibration increase in 1X and 2X components was much faster than expected and documented in the literature so far.
- "Split amplitude" characteristics through critical speeds.
- The presence of a 2X component at steady state operation (very small) and run-up and run-down (considerably higher).
- Because of the 2X excitation a peak will be seen already at half of its actual speed value considering the unfiltered overall vibrations.
- The 1X critical speeds will drop because of stiffness loss.
- To localize the axial crack position, the vibration amplitude might be misleading. The phase behavior during run-up and run-down will give good information. At the spot where the phase will not change from 1st to 2nd critical, there the excitation occurs.

3.13.3 MW Steam Turbine, Lateral Crack

In Fig. 3.189, the cross-section drawing of the turbine is shown.

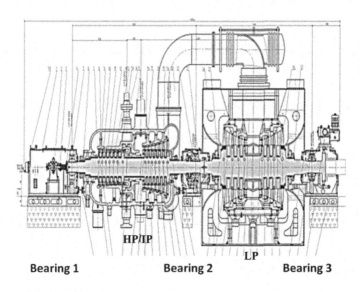

Bearing 1   **Bearing 2**   **Bearing 3**

Fig. 3.189 Steam turbine

Figure 3.189 shows a drawing of a Chinese turbo-set consisting of a combined high-pressure–intermediate-pressure stage (HP/IP) and a low-pressure stage (LP).

At about midnight on the 5th of March, there was a vibration increase at all the bearings, pronounced at bearing #3, after a long period of stationary vibrations at full load.

Figure 3.190 documents this vibration increase. After shutting down, we also see elevated vibrations at turning gear. The striking fact is the extremely high vibrations at turning gear, only 5 rpm.

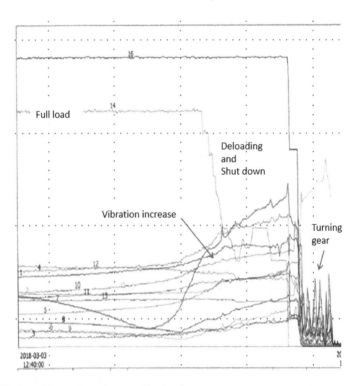

Fig. 3.190 Surveillance chart shows a vibration increase

The run-down on the 5th of March (see Fig. 3.191) indicates a second new peak at the lower blue line (overall vibration component), but not at the higher red line (1X unbalance component), which means that the frequency of this peak is not rotational 1X. This second peak is exactly at half of the 1800 rpm critical speed at around 900 rpm, excited by 1X unbalance. The next page will show that the peak at half of the critical has the twice rotational frequency 2X. There is now a mechanism in the shaft, which produces 2X vibrations. This is a new phenomenon, which was not there before. This can be seen in Fig. 3.191: This earlier measurement shows that there was no 2X frequency noticeable (Fig. 3.192).

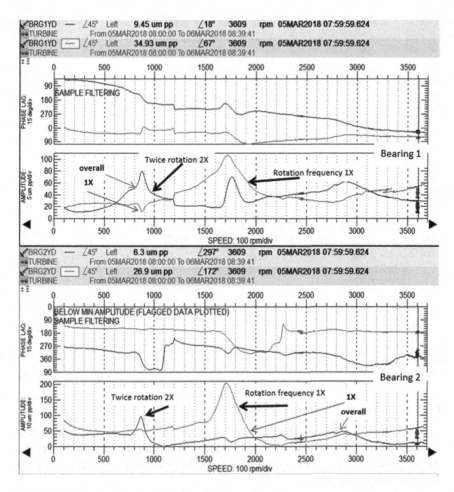

Fig. 3.191 Indicates vibrations during run-down

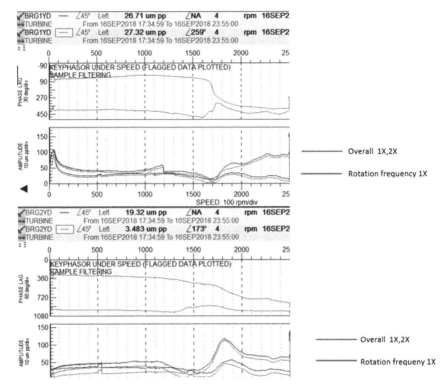

Fig. 3.192 An earlier Bode plot from January

When the speed drops to half of the critical, this mechanism excites the critical again, because its frequency is 2X, it has twice the rotation frequency. We see this 2X phenomenon also at the orbit (Fig. 3.193). The shaft orbit performs two loops now because of the presence of 2X.

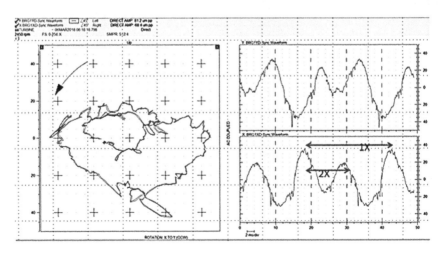

Fig. 3.193 The orbit and time function

The frequency spectrum at the speed of the peak shows a dominating 2X frequency (see Fig. 3.194), though the shaft is rotating with half of this frequency. We did not allow to run up the unit again and insisted on a rotor check.

We suspected a crack in the turbine rotor. Figure 3.195 shows the drawing of the rotor with the crack area, and in Figs. 3.196 and 3.197, we can see the cracks in the IP-LP coupling.

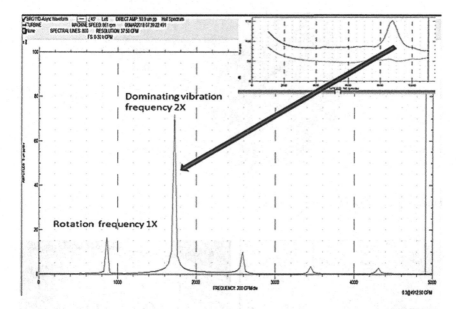

Fig. 3.194 Frequency spectrum

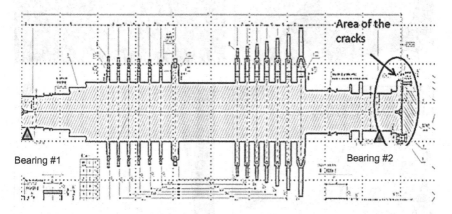

Fig. 3.195 Cross section of the turbine rotor

The area of the ellipse indicates the coupling to the LP, where the cracks happened.

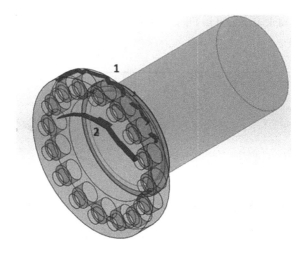

Fig. 3.196 Location of the crack

Fig. 3.197 Photos of the coupling cracks

As can be seen in Fig. 3.198, we also found microcracks in the coupling holes, due to sloppy assembly and disassembly from which we think some of the cracks seen in Fig. 3.197 originate.

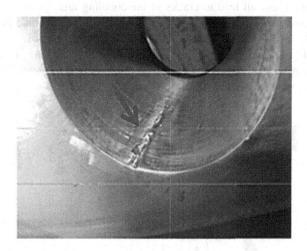

Fig. 3.198 Microcracks in the coupling holes

In addition to the crack, bearing #2 had been damaged badly due to overloading (see Fig. 3.199).

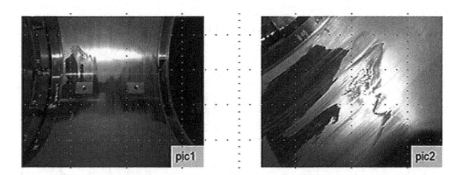

Fig. 3.199 Damage of bearing #2

Various explanations for the crack root cause had been circulating, and our opinion about the crack origin is the following:

The alignment according to Fig. 3.200 was one of the reasons for the cracks. The key issue is the bad cold alignment, shown in the figure.

There is the bottom gap of the coupling of 0.33 mm and additionally the lift of bearing #2 by 0.24 mm. When forcing the coupling halves together by means of the screws, an enormous bending stress will be created, because bearing #2 is very close. In addition, the thermal growth of pedestal #2 will have an effect when the unit heats up. These all lead to cracks in the coupling area, starting at the micro-cracks caused by the sloppy assembly. Additionally, bearing #2 had been damaged by overload.

The HP-IP rotor had been repaired at a big machinery company in Indonesia by manufacturing a new IP coupling end and welding it on. Figure 3.201 shows the repair of the rotor by welding on a new coupling end.

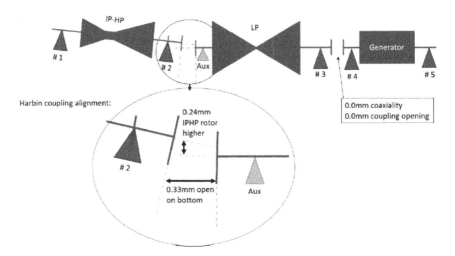

Fig. 3.200 Alignment of HP-IP turbine regarding LP

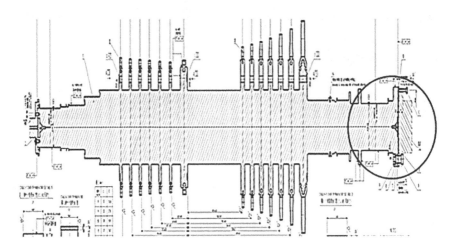

Fig. 3.201 Rotor with new welded on coupling end

A new coupling end had been welded on, indicated by the circle (see Fig. 3.201).

The rotor had been installed with a new alignment without vertical offset and no bottom opening of the coupling. Table 3.7 shows the measures of the new alignment.

Table 3.7 Gaps of new alignment		Offset (mm)	Gap (mm)
	HP-IP to LP Coupling	0.00	0.10 (open at the top)

3.13.4   MW Combined Cycle Plant, Ring-Shaped Cracks

In contrary to the lateral crack, there is no 2X vibration at a ring shaped cracked, because the ring crack does not cause any uneven polar stiffness of the cross section and therefore no sag changes during rotation (Fig. 3.202).

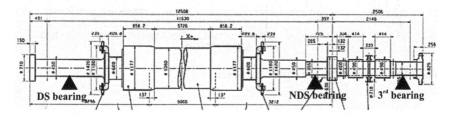

Fig. 3.202   400 MW generator rotor of the turbo-set

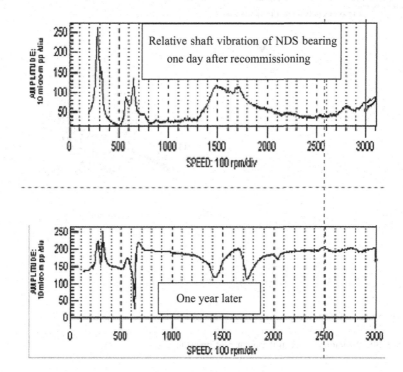

Fig. 3.203   Measurements of the run-down vibrations

Figure 3.203 shows data taken from a combined cycle plant in Spain. The upper graph shows the run-up vibrations of the NDS generator bearing after an overhaul. This was the normal and expected vibration signature. The lower graph shows the same after one year of operation. The vibrations are elevated to a level of approx. 170 $\mu m_{p.p.}$.

The vibration pattern now looked like the rotor suffers from a heavy run-out fault of about 170 μm (see Fig. 3.204).

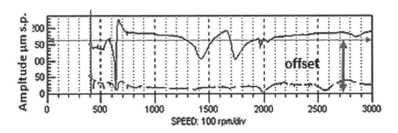

Fig. 3.204 Speed-independent offset of vibrations

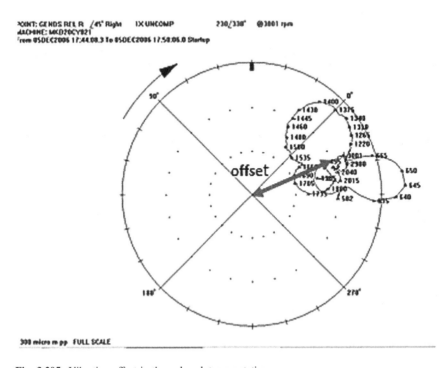

Fig. 3.205 Vibration offset in the polar plot presentation

Figure 3.205 shows a considerable run-out error (compare with fair run-out in Fig. 3.114). The polar plot loops, standing for the critical speeds, show an offset of the center of the graph. The amount of the offset indicates the amount of the run-out error.

Checking also the polar plot in Fig. 3.205, we see the typical behavior of a shaft flawed with a run-out fault. The origin of the plot shows an offset from the center having approx. the amount of the 170 μm run-out fault.

Surprisingly, the measurement of the run-out (with dial indicators, at 0 rpm) at the slip-ring shaft did not indicate a run-out error (only 0.02 mm eccentricity was measured). During the run-down, there is an "offset" covering the entire speed range around 170 $\mu m_{p.p.}$. This is the signature of a mechanical run-out (eccentricity-) error, which is a geometrical offset of the shaft, not being directly a function of the speed like an unbalance.

An unbalance develops with speed according to a square function:

$$F = mr\omega^2 \qquad (3.1)$$

where:

F = centrifugal force, m = rotating mass, ω = frequency or speed, r = radius.

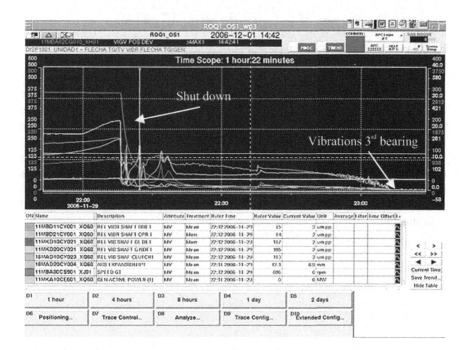

Fig. 3.206 Vibrations during run-down after trip

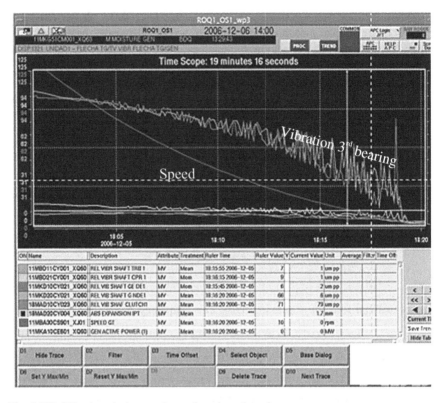

Fig. 3.207 Vibrations during run-down after trip, enlarged

The conclusion looking at Figs. 3.206 and 3.207 was that there is a crack in the slip-ring shaft, whereby the centrifugal force holds the shaft in an offset position from approx. 80 rpm upward due to centrifugal force. The shaft behaves then according to a heavy run-out error.

At speeds lower than 80 rpm, the centrifugal force is too low, and the shaft starts to kink within the tolerance, given by the crack (see Fig. 3.208). At very low speeds and standstill, the spring forces of the uncracked material bring the shaft back into the original position and therefore we do not measure the run-out fault with the dial indicators.

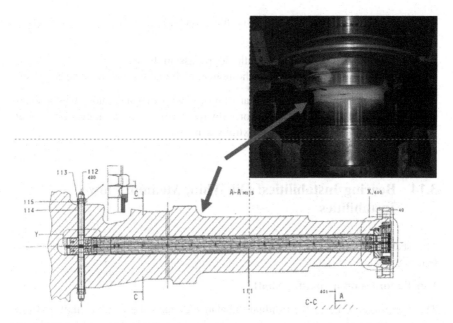

Fig. 3.208 Location of the crack

There was a ring-shaped crack not in one plane, but axially orientated like a "spiral" (see Fig. 3.209). If the crack would not have been "spiral like," the shaft would have disintegrated.

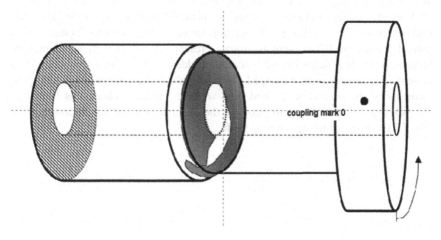

Fig. 3.209 Ring crack

This is an example how a ring-shaped crack can be detected by means of the vibration signature despite there is no other dominating frequency than 1X.

The cracks presented here had three main causes:

• inadequate design,
• bad alignment in combination with sloppy assembly and
• rotor material had been made in-homogenous by additional welding at jobsite.

A mechanical run-out measured at standstill with dial indicators was not necessarily an indication for a good rotor. Always verify your dial indicator run-outs with the slow roll signals of the proximity sensors.

3.14   Bearing Instabilities: Oil Whip, Medium Flow Instabilities

This section describes the forces acting on a rotating shaft at a certain point in time (snapshot).

Kinetic forces on a rotating shaft:

The centrifugal force of the residual unbalance deflects the flexible shaft, until the elastic restoring force, due to the shaft deflection, is being built up until equilibrium. The oil film action of the bearings delivers an external damping force. An internal mechanism can create a destabilizing force, which counteracts to the damping force and can make the shaft instable.

Self-excited vibrations can be recognized by the fact that their frequency generally corresponds to that of the weakest damped natural frequency in the system. With shafts rotating above the critical speed, this is almost always the first critical. At first, an external influence excites the natural-frequency vibration. Energy is removed from the rotating shaft and transferred to the vibrating system, at the natural frequency. If the damping forces present are not enough to counter this process, a steadily increasing vibration takes place, at the natural frequency of the system. The increase is not stopped until the end of the linear range of the spring elements of the system is reached, i.e., when the bearing clearances have been exhausted or the rotor starts to graze on the housing. The kinetic forces shown in Fig. 3.210 are fixed to the rotating shaft.

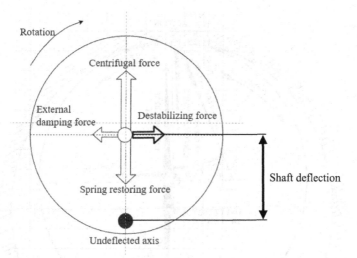

Fig. 3.210 Kinetic forces acting on a rotating shaft

Generally, self-excited vibration can be corrected by the following measures:

- raising the external damping (bearing damping and damping of the external structure) and
- raising the critical speeds of the rotor.

The practical experience we made has shown that two types of bearing instabilities occur:

- The high specific bearing load may cause a temporary metal contact of the journal and may lead to a friction whirl.
- If a bearing has an extremely low radial specific bearing load, it may develop an oil whip, which can end up in an oil whip in the case of machines with over-critical rotors.

Friction Whirl:

At a highly radial loaded bearing, a mixed friction between fluid and dry can develop. A tangential Coulomb friction force will then be induced, compensating the external damping force. This can be the precondition for instability.

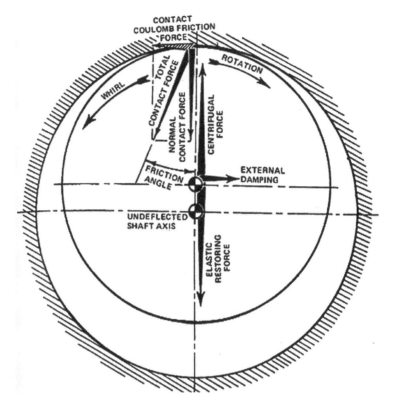

Fig. 3.211 Forces at a friction whirl (Harris 1961)

When the friction whirl (see Fig. 3.211) occurs, the shaft whirl turns against its rotation sense. The vibration frequency will correspond to the 1st critical of the shaft, despite it rotates with higher rotational frequency (at overcritical elastically behaving shafts).

Oil whip:

If the bearing has a very low specific radial bearing load, the consequence could be an oil whip.

As shown in Fig. 3.212, the rotor is supported by a thin film of oil. The entrained fluid circulates at about half the speed of the rotor (the average of shaft and housing speeds). Because of viscous losses in the fluid, the pressure ahead of the point of minimum clearance is lower than behind it. This pressure differential causes a tangential destabilizing force in the direction of the rotation that results in a whirl— or precession of the rotor at slightly less than half rotational speed (usually 0.43–0.48 times the rotational speed).

Considering a run-up, before the 1st critical speed of the rotor is reached, the shaft will vibrate with this approximate half of the rotational frequency (because the circumference oil film speed is approx. half of the shaft rotation). As soon as this ½ rotational component joins into the 1st critical, the frequency of the rotational component remains at the frequency of the 1st critical, independent of the rotor speed (see Fig. 3.213). The amplitude of the vibrations will considerably elevate in an erratic way.

Whirl is inherently unstable, since it increases centrifugal forces which in turn increase whirl forces. The rotation energy drives the instable system and it is often limited only by geometrical properties. Stability is normally maintained through external damping in the rotor-bearing system.

Lube oil pressure forces can initiate an oil whip (see Fig. 3.212). The spectrum, when the rotor runs into an oil whip, is shown in Fig. 3.213.

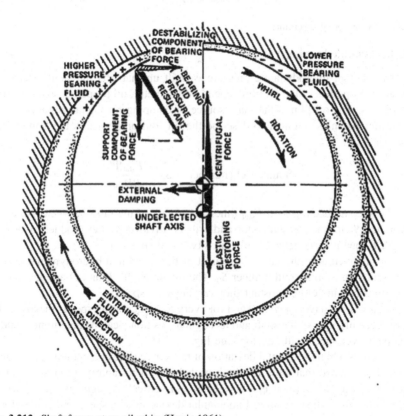

Fig. 3.212 Shaft forces at an oil whip (Harris 1961)

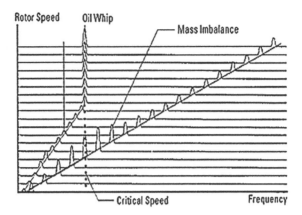

Fig. 3.213 Waterfall spectrum

Hydrodynamic bearings:

Under rotation, hydrodynamic pressure is built up that lifts the shaft. Static and dynamic properties of the oil film depend on geometry, oil properties and operating conditions. The oil film is the main source of damping for lateral vibrations.

A dimensionless number to describe the oil film properties is the Sommerfeld number S_0:

$$\text{Sommerfeld number:} \quad S_0 = \frac{F_{\text{stat}}\Psi^2}{BD\eta\omega} \tag{3.2}$$

The specific bearing load F_{stat} times the square of the ratio between radius and clearance Ψ forms the numerator and will be divided by the product of bearing length B, bearing diameter D, oil viscosity η and speed ω.

The static equilibrium curve in respect of the speed in a hydrodynamic bearing represents the Sommerfeld number S_0. The quantities Ψ, B and D are geometrical quantities and therefore constant (not varying with speed).

At $S_0 = 1$ we have *speed* = 0 and for every S_0 at increasing speeds the according to the eccentricity ε of the shaft and its angular position γ can be determined, and so this path over speed will develop (see Fig. 3.214).

If there is a high horizontal deviation of this curve (like at a cylindrical bearing), it can be assumed that there is a high probability of instability, because of strong cross-coupled stiffness between horizontal and vertical. A segment bearing, *like a tilting bearing*, will show almost no angular deviation with speed and is therefore a very stable operating bearing.

Fig. 3.214 Course of the Sommerfeld number over the speed

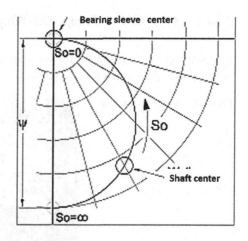

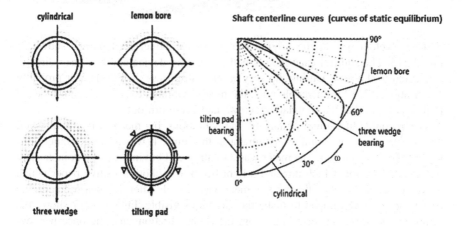

Fig. 3.215 Static equilibrium curves for different bearing types

According to Fig. 3.215:

- Cylindrical bearings are most responsive for instabilities.
- Tilting pad bearings have practically no instability tendency.
- Three wedge bearings are better than lemon bore regarding the stabile operation of the shaft.

3.14.1 Feed Water Pump, Oil Whip

This was our first assignment for a vibration problem abroad *in 1970*. It was the harbor power plant in Benghazi, Libya.

Fig. 3.216 Boiler feed pump

The problem was at the driving motors of the boiler feed pumps shown in Fig. 3.216. The squirrel-cage motors had no starting device and had been directly connected to the grid. During the very short run-up (less than 2 s), after probably 1 s or about 47–50 rotor revolutions, the motors vibrated excessively and blocked in the stator so that the foundation screws had been torn out.

We recorded the time signals of the vibrations by means of a Honeywell "Visicorder," a very common instrument at that time. It was a multichannel voltmeter (max. 6 channels). The meter was a rotating coil but instead of a pointer there was a small mirror, which directed a light beam to a photo-sensitive paper which moved continually when the measurement was started. After the measurement, the paper had to be developed to make the vibrations visible. Figure 3.217 shows the setup for the vibration recordings from the shaft vibrations and the once per revolution signal as a speed reference.

Fig. 3.217 Setup for vibration measurements

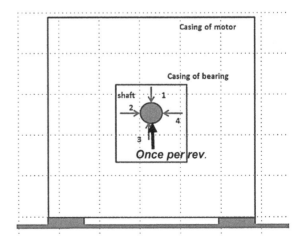

Figure 3.217 shows the setup for the vibration recordings from the shaft vibrations and the once per revolution signal as a speed reference.

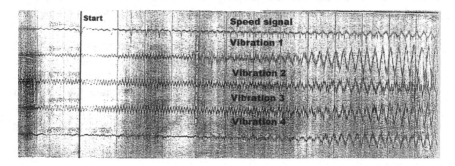

Fig. 3.218 Visicorder recording of the motor shaft vibrations

Figure 3.218 shows a recording was made 1970 by means of a Honeywell "Visicorder," a very common instrument at that time. It was a multichannel voltmeter (max. 6 channels). The meter was a rotating coil but instead of a pointer there was a small mirror, which directed a light beam to a photo-sensitive paper which moved continually when the measurement was started. After the measurement, the paper had to be developed to make the vibrations visible.

The original concept of the rotor was under-critical, but the Visicorder recordings displayed in Fig. 3.218 made clear that the motor was over-critical and had to pass a critical speed before reaching nominal speed.

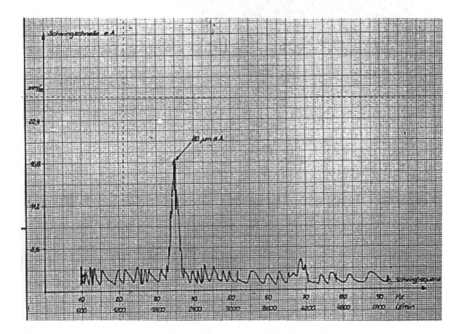

Fig. 3.219 Evaluation of the tape recording

We tried to identify the critical speed of the rotor by means of a magnetic tape recording (see Fig. 3.219). We made an endless loop of magnetic tape of the run-up vibrations and analyzed them with a tuning filter. The loop ran continuously at 10 times the tape speed. We found the 1st rotor-critical at 34 Hz. This confirmed the findings of the Visicorder measurements.

In Fig. 3.220, we see an enlargement of the recording. This snapshot was taken approx. 1 s after switching the motor to the grid. We see that the period time of the vibration signals is longer than the duration of 1 revolution. So, the vibration frequency after 10 s of run-up is already lower than the rotation frequency. In Fig. 3.220, the rotor rotates with approx. 49 Hz (2940 rpm) and its vibration frequency is about 34 Hz. The ring lubricated bearing had a very low specific bearing load of little bit over 0.1 MPa.

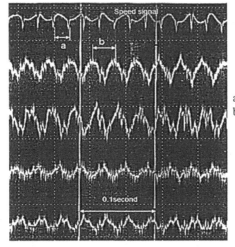

a: period of rotation
b: period duration of vibration

Fig. 3.220 Enlarged Visicorder recording

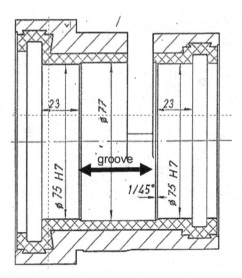

Fig. 3.221 Drawing of modified bearing

We increased the bearing load by machining a groove according into the babbitt material according to the drawing displayed in Fig. 3.221. We converted a low loaded bearing into two higher loaded bearings which cured this problem.

3.14.2   50 MW, 60 Hz, Combined Cycle Plant, Friction Whirl

Heavy relative shaft vibrations occurred at the front bearing #1 of the gas turbine with a frequency of 17–18 Hz, which corresponded to the 1st critical of the turbine rotor. The bearing arrangement is shown in Fig. 3.222.

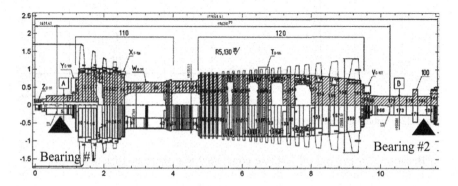

Fig. 3.222 Gas turbine bearing arrangement

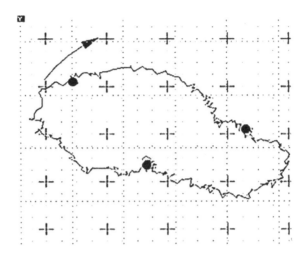

Fig. 3.223 Measured vibration orbit

In the vibration orbit (see Fig. 3.223), there are 3 key-phasor marks (black dots), which means during a full orbit, the shaft makes 3 revolutions. So, the vibration frequency is about 3 times lower than the rotating speed which is 3600 rpm *at 18 Hz* (Fig. 3.224).

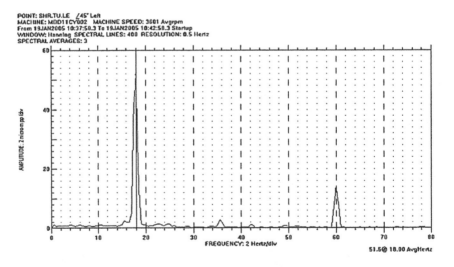

Fig. 3.224 Vibration spectrum of the instable rotating shaft

Because of a jacking oil test, we concluded that there is a friction whirl.

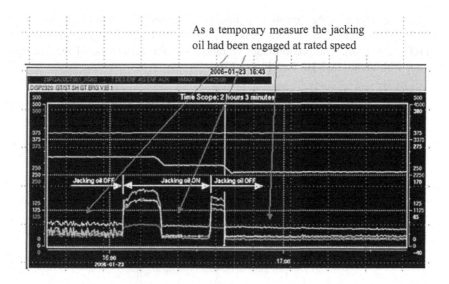

Fig. 3.225 Surveillance recording during the jacking oil test

The engagement of the jacking cured the shaft instability instantly according to Fig. 3.225. The calculated oil flow at the bearing was 4.5 l/s, and the measured oil flow was 2.2 l/s. This had been corrected by a modified oil pump concept, which cured the problem permanently.

3.14.3 80 MW Air-Cooled Generator of a Gas Turbine

At the driven side bearing (see Fig. 3.226) of an air-cooled generator of a 80 MW GT turbo-set in Azerbaijan, shaft vibration suddenly jumped up and tripped the unit (see Fig. 3.227).

Fig. 3.226 Driven side bearing (DS) of the generator

The spectrum at the vibration highpoint before the trip showed a predominant 18.5 Hz. (GT speed: 6300 rpm, generator speed: 3000 rpm, 1st generator critical: 1100–1200 rpm). The predominant vibration was the 1st critical of the generator rotor with about 18.5 Hz (see Fig. 3.228).

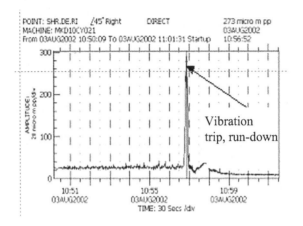

Fig. 3.227 Time trend at vibration jump

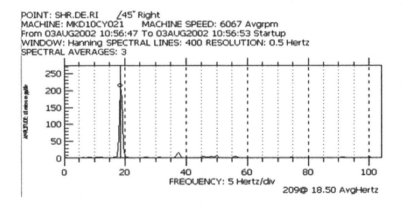

Fig. 3.228 Spectrum at the high point of the vibrations

Figure 3.227 shows a sudden vibration increase in the DS bearing during normal operation (black arrow). This sudden increase in vibrations tripped the unit.

Figure 3.229 shows a full vibration period, indicated by the closed orbit. The 6 dots indicate the key-phasor marks at the gas turbine. Because the turbine is running with approx. double the speed than the generator (gas turbine speed: 6070 rpm; generator speed: 3000 rpm), one would normally expect two key-phasor markings (dots) during a full vibration period.

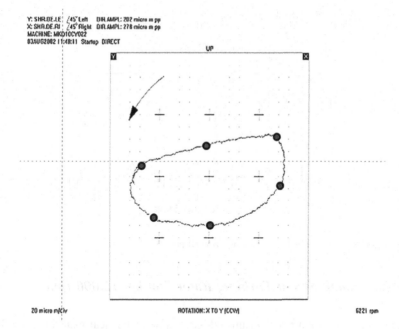

Fig. 3.229 Orbit at the DS generator bearing

Because we see 6 key-phasor markings in Fig. 3.229, we can conclude that the orbit of the generator DS bearing proves a vibration frequency being approx. 6 times lower than the rotational frequency of the GT or approx. 3 times lower than the rotational frequency of the generator. According to the spectrum, the relevant vibration frequency is 18.5 Hz, which is approx. 3 times lower than the rotational frequency of the generator with 50 Hz. These 18.5 Hz correspond approx. to the 1st critical speed of the generator rotor and can be identified as instability.

It was found that the DE bearing of the generator was loaded too lightly. As an improving measure, the bearing pedestal had been lifted by 0.2 mm, which cured the problem. Figure 3.229 shows the spectrum after lifting the pedestal.

The lifting of driven side generator pedestal resulted in a very fair vibration behavior according to Fig. 3.230.

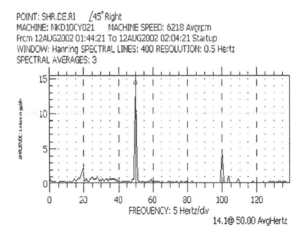

POINT: SHR.DE.R1   ∠45° Right
MACHINE: MKD10CY021   MACHINE SPEED: 6218 Avgrpm
From 12AUG2002 01:44:21 To 12AUG2002 02:04:21 Startup
WINDOW: Hanning SPECTRAL LINES: 400 RESOLUTION: 0.5 Hertz
SPECTRAL AVERAGES: 3

FREQUENCY: 5 Hertz/div

14.1@ 50.00 AvgHertz

Fig. 3.230 Spectrum after lifting bearing pedestal DS

3.14.4 Small Steam Turbine, Rotor 760 kg, 12,000 rpm

Figure 3.231 shows a small steam turbine in a petrochemical plant in Israel. It is used for driving an ethylene compressor. The weight of the rotor is 760 kg and the nominal speed is 12,000 rpm. The bearing distance had been 1865 mm.

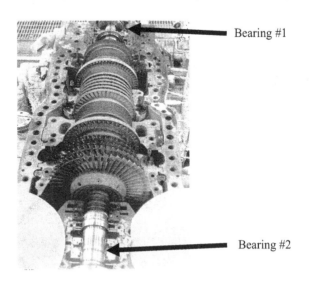

Bearing #1

Bearing #2

Fig. 3.231 Picture of the open turbine

We had been called to rebalance the turbine rotor, because high vibrations had been noticed at the front bearing #1. When we arrived, the unit was running. But the running noise was somehow strange and sounded like a rumbling. Also, when we touched the front bearing one could feel a rumbling or rattling. A machine having only an imbalance sounds and feels much more harmonic. So, at first a frequency analysis was carried out.

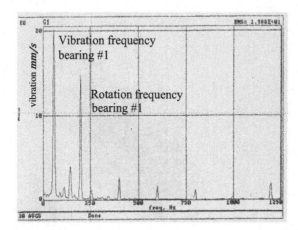

Fig. 3.232 Spectrum before any action

The spectrum in Fig. 3.232 was taken before we started to investigate the bearing #1. The dominating vibration frequency according to Fig. 3.232 was not the rotation frequency of 200 Hz (as it should be in an unbalance case), but a much lower component of 55 Hz was the dominating frequency, so we concluded a bearing instability.

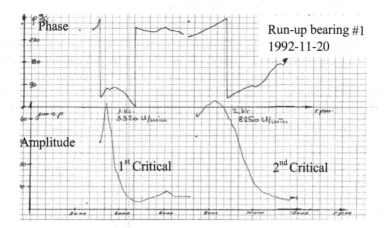

Fig. 3.233 Tape recording of the run-up

Figure 3.233 shows the graph of an old X/Y recorder, taken from the tape recording. The vibration signal had been recorded on a magnetic tape in order to determine the shaft criticals. The 1st critical can be seen at 3320 rpm, and the 2nd critical is at 8250 rpm.

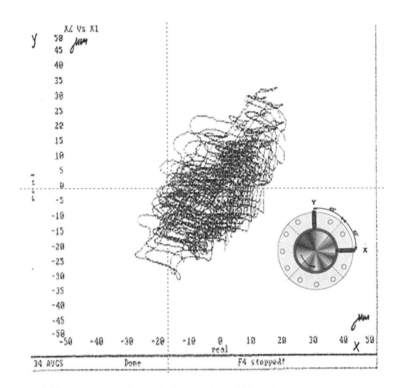

Fig. 3.234 Orbits over 34 revolutions before any improving action

We also recorded the orbit of the shaft of bearing #1 from the tape recording by means of a 2-channel analyzer from Zonic. In Fig. 3.234, you can see the orbits of about 34 revolutions and the shaft is bouncing around the bearing sleeve. After having stopped the unit, it was cooled down and the bearing #1 had been opened.

Both bearing halves had been already damaged considerably, but the shaft was still in order (see Fig. 3.235).

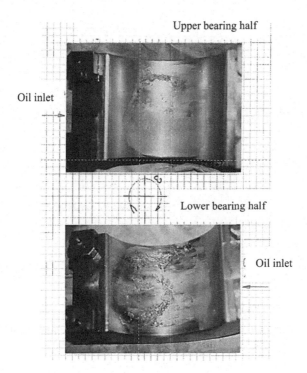

Fig. 3.235 Considerable damage of bearing #1

The bearing halves had been casted with new Babbitt material, but the specific bearing load had been increased from approx. 0.5 to about 1.0 MPa. This was done by reducing the axial length of the bearing by half.

In Fig. 3.236, we see that after the bearing repair the dominating frequency was the rotation frequency. The vibrations with the frequency of the 1st critical had been marginally small with about 1.5 mm/s. The bearing modification cured the instable behavior of the shaft.

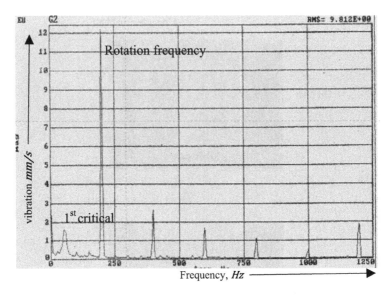

Fig. 3.236 Spectrum after bearing modification

Figure 3.237 shows the orbits of 35 revolutions after the modification. The shaft now performs a stable rotation.

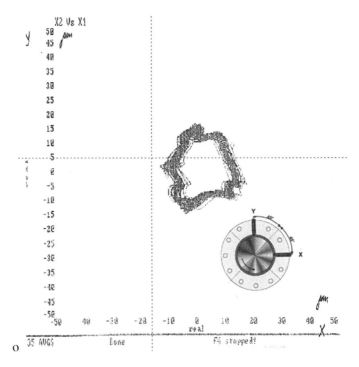

Fig. 3.237 Orbits after bearing modification

3.14.5 Medium Flow Instability

According to Fig. 3.238, the residual unbalance is creating an unsymmetrical rotor fixed gap of the rotor blading against the stator. Here, we have the wide gap on top, which causes a reduced driving force. At the smaller bottom gap, more steam can flow through the blades and there is a higher driving force. Consequently, a perpendicular compensation force develops. This force may trigger an instable running of the rotor (Figs. 3.239 and 3.240).

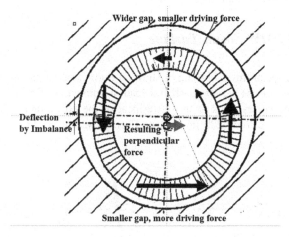

Fig. 3.238 Forces acting on the rotating shaft

Fig. 3.239 135 MW steam turbo-set at the Philippines

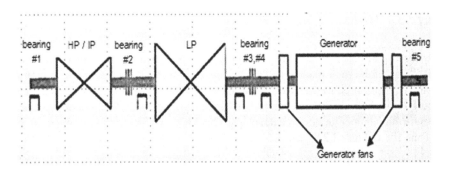

Fig. 3.240 Turbine and bearing arrangement

3.14.6 135 MW Steam Turbine

The waterfall spectrum in Fig. 3.241 shows a normal vibration behavior in stable condition. As soon as we approach full load, a dominating frequency component appears, which corresponds to the 1st critical of the HP rotor (28.7 Hz) and can reach amplitudes of over 200 $\mu m_{p.p.}$ (see Figs. 3.242 and 3.243).

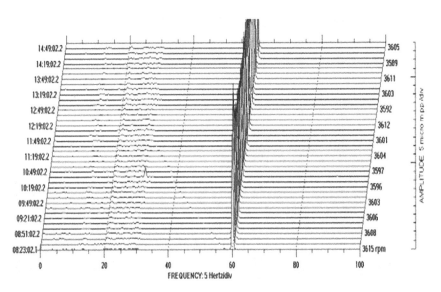

Fig. 3.241 Waterfall spectrum at max.68 MW from bearing #1

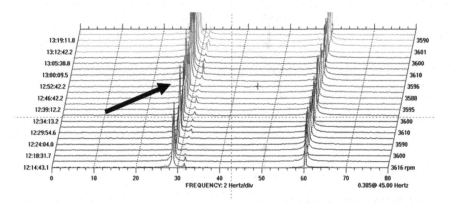

Fig. 3.242 Waterfall spectrum at full load, bearing #1

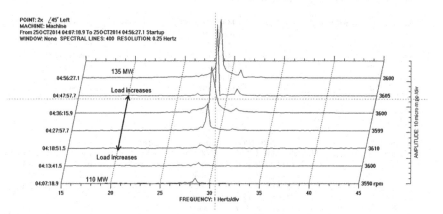

Fig. 3.243 Waterfall at the upper end of the load range

The instability develops at the end of the load range as can be seen in Fig. 3.243. Between 110 and 135 MW, a fast vibration increase in the 28 Hz component takes place. The vibrations jump from 80 to 90 $\mu m_{p.p.}$ up to over 200 $\mu m_{p.p.}$ and the dominating vibration frequency changes from 60 Hz to around 28 Hz.

This instability can be reduced by a certain opening concept of the HP governor valves (see Fig. 3.244): The upper valves should stay more open than the lower ones, expecting that the steam forces from above will dominate and the rotor will be pressed into its bearings.

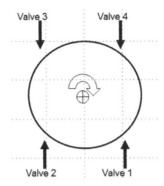

Fig. 3.244 Governor valve arrangement

Table 3.8 Shaft vibrations of the turbo-set

Ch #	Channel Name	Machine Name	Amplitude Units
1	1x	Machine	micro m pp
2	1y	Machine	micro m pp
3	2x	Machine	micro m pp
4	2y	Machine	micro m pp
5	3x	Machine	micro m pp
6	3y	Machine	micro m pp
7	4x	Machine	micro m pp
8	5x	Machine	micro m pp

Sample 17

Channel	Date/Time	Speed	Direct	Gap	1X
1	27 OCT2014 09:36:27.7	3600	55.7	-9.78	29.0
2	27 OCT2014 09:36:27.7	3600	79.5	-10.9	40.1
3	27 OCT2014 09:36:27.7	3600	177	-12.0	72.6
4	27 OCT2014 09:36:27.7	3600	117	-13.3	33.3
5	27 OCT2014 09:36:27.7	3600	70.5	-9.89	44.3
6	27 OCT2014 09:36:27.7	3600	40.1	-10.9	13.1
7	27 OCT2014 09:36:27.7	3600	98.2	-10.1	91.4
8	27 OCT2014 09:36:27.7	3600	72.7	-9.08	65.4

According to the readings in Table 3.8, the high spots are at bearing #1, #2 and bearing #3. The difference between "Direct" and "1X" indicates the presence of a non-rotational frequency component *with the frequency of the 1st critical of HP/IP rotor identified as an instability.*

The countermeasures for this problem were:

- increase the bearing load at bearing #2 by lowering bearing #1 by 0.2 mm and lift bearing #2 by 0.2 mm,
- shorten the axial length of babbitt metal of bearing #2 by 10 mm each side and
- decrease the vertical bearing clearance from 0.45 mm to 0.39 mm.

By means of these measures, the instability could be cured.

3.14.7 Instable Compressor Air Flow, Rotating Stall

During a hot run–down, a pronounced vibration peak occurred at around 1500 rpm with 500 $\mu m_{p.p.}$ amplitude shaft vibration for a few seconds. This phenomenon is usually called a rotating stall.

The speed trend (run-down) displayed in Fig. 3.245 shows that the shaft becomes instable at around 1500 rpm for a short time.

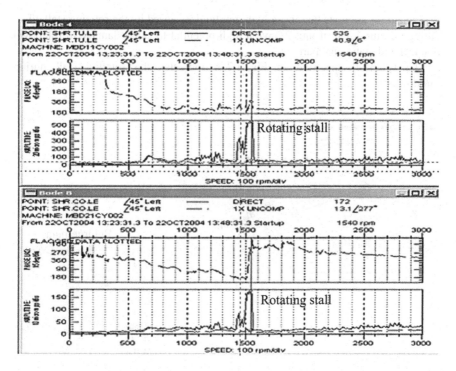

Fig. 3.245 Speed trend

Figure 3.245 displays the speed trends of the run-down measured at the turbine bearing (top) and the compressor bearing (bottom). It shows that the shaft becomes instable at around 1500 rpm for a short time, but with very high vibrations.

From a report of the engineer from jobsite:

> ... the 12 Hz-component, which is present especially on the waterfall plots of the relative shaft vibration at around 1500–1600 rpm ..., was caused by the rotating stall effect, a flow separation phenomenon. It is not possible to reduce this component by balancing. During the shutdown of the 22 October, the rotating stall effect caused very high relative amplitudes of over 550 $\mu m_{p.p.}$ (!) on the measuring spot turbine front bearing ...

The author assumes that those high amplitudes caused by the rotating stall effect during the run-down of unit 2 have a high potential to cause damage to rotational (blades) and stationary parts.

The waterfall spectrum in Fig. 3.246 indicates the rotating stall during run-down.

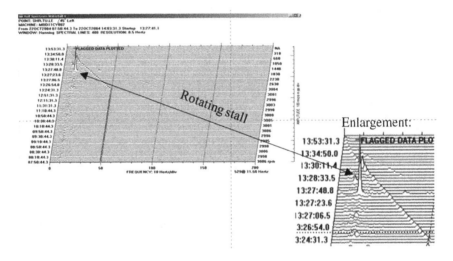

Fig. 3.246 Waterfall spectrum indicates rotating stall

At the waterfall spectrum in Fig. 3.246, we see a sharp vibration peak at around 1500 rpm during the run-down of the warm turbine (black arrow).

Figure 3.247 shows a recording of a provoked rotating stall at our test center. At compressor blade #8, the pressure had been recorded and at the same time the shaft vibrations at the compressor bearing.

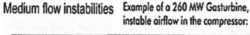

Medium flow instabilities Example of a 260 MW Gasturbine, instable airflow in the compressor:

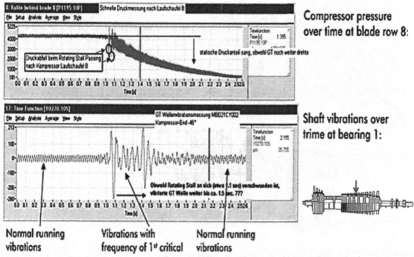

Compressor pressure over time at blade row 8:

Shaft vibrations over trime at bearing 1:

| Normal running vibrations | Vibrations with frequency of 1st critical | Normal running vibrations |

Airflow instability indicated by pressure drops (being superimposed over static pressure) excite rotor vibration at 1st critical frequency

Fig. 3.247 Shaft vibrations together with the pressure measurements in row 8 of the compressor

If the unit is running, we see a certain static pressure as well as the low shaft vibrations with rotation frequency. When the unit is shut down, sudden pressure drops occur and at the same time, the rotor starts to vibrate excessively with its 1st critical. In the meantime, the speed has dropped, the pressure fluctuations disappear and the vibrations decay and return to rotation frequency until we have again a low level. This process lasted about 1 s.

In Fig. 3.248, the effect of the blow off valves is shown. The gas turbine of our test stand was running at 1200 rpm and shut down with open and closed blow off valves. When the blow off valves were closed the air flow in the compressor was blocked, which causes flow perturbations exciting the 1st rotor critical.

Our practice showed that:

- An instability caused by the medium flow excites the 1st critical of the rotor as well as the bearing instabilities.
- The oil whip in bearings and the media flow instabilities are connected. Media flow instabilities can be influenced by bearing measures and other way round.

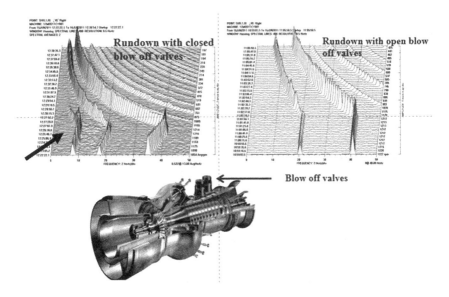

Fig. 3.248 Effect of the blow off valves at a gas turbine

Figure 3.248 shows two waterfall spectra of different run-downs. The left spectrum shows a run-down with closed blow-off valves, leading to subharmonic frequencies in the spectrum (indicating a rotating stall). The right spectrum shows a run-down with open blow off valves, where the subharmonics are not existent. Therefore, an effective measure to avoid a rotating stall will be the opening of the blow off valves during run-down of the warm gas turbine.

Chapter 4
Jobsite Balancing

Jobsite balancing cannot be considered as a balancing method in order to correct the mass unbalances of the rotors (with the exception of a re-blading action during an overhaul jobsite). This has already been done in the factory of the rotors before delivery. Jobsite balancing deals with all the influences, which the rotors will be exposed to at jobsite. These are:

- influences of coupling together,
- influences of temperature,
- influences of torque,
- influences of alignment and
- influences of other parameters like oil temperature, vacuum, valve sequence, etc.

The improvement of the vibration impacts of these influences by additional balance weights is what we call the jobsite balancing.

Procedures in which the method of correcting unbalances is performed in one or more correction planes are recommended if:

- Rotors are supported in several bearings and having several balancing planes, e.g., see the shaft model shown in Fig. 4.1 with bearings 1, 2, 3 and 4 and balancing planes A, B, C and D.
- Rotors are flexible: we consider a rotor then flexible, when its first critical speed is below the operation speed.

© Springer Nature Switzerland AG 2020
F. Herz and R. Nordmann, *Vibrations of Power Plant Machines*,
https://doi.org/10.1007/978-3-030-37344-3_4

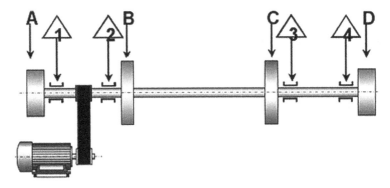

Fig. 4.1 Shaft with 4 balance planes (Brüel and Kjäer Vibro 1995)

Table 4.1 shows assumed measuring values for the balancing demonstration:

  i. The vibrations at the original unbalance condition are measured (date or reference run).
 ii. A test mass is attached in plane A close to bearing #1. The rotor is started again, and after having measured the vibrations, the weight will be removed.
iii. The test weight is attached in plane B, vibrations are measured, and the weight is removed.

After having covered all 4 balance planes in the same way, we have obtained all the possible weight influences (influence vectors) which is the basis for the actual balance weight or weight combination (see Fig. 4.2).

For every measured bearing, a graph will be obtained and therefore we will get an influence vector for the corresponding weight for every bearing. The connection to the reference run ("a" in Fig. 4.2) is called the influence vector.

Table 4.1 Measuring values

Run No.	Remarks	Unbalance Vibrations at bearings close to the planes:							
		Bearing1		Bearing2		Bearing3		Bearing4	
		Ampl.	Angle	Ampl.	Angle	Ampl.	Angle	Ampl.	Angle
1	Original Unbalance	38,0	105°	25,5	126°	21,0	243°	34,0	276°
2	Test mass m = 350 g at angle 0° applied in Plane A	33,0	31°	16,0	103°	32,5	220°	31,0	270°
3	Test mass m = 350 g at angle 0° applied to Plane B	41,5	72°	34,0	47°	19,0	262°	33,0	292°
4	Test mass m = 350 g at angle 0° applied to Plane C	47,8	100°	29,5	76°	25,3	184°	28,2	244°
5	Test mass m = 350 g At angle 0° applied to Plane D	26,4	98°	24,6	92°	22,4	235°	18,5	165°

Fig. 4.2 Measured values plotted into a polar plot (Brüel and Kjäer Vibro 1995)

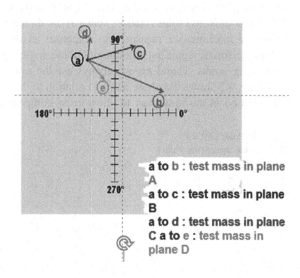

a to b : test mass in plane A
a to c : test mass in plane B
a to d : test mass in plane C a to e : test mass in plane D

We do balancing of rotors under operational conditions at jobsite. This process does not only correct mass distribution failures of the rotor; it generally reduces vibration (with rotational frequency) by means of the aid of balancing weights. These vibrations might be caused by various effects or sources such as mechanical run-outs, temporary rotor bending (thermal unbalances), etc.

When a machine is in operation, especially an overcritical rotor, the vibration values never are constant or equal throughout all the operating conditions. We will encounter a transient vibration behavior, being governed by certain operating parameters (temperature, torque, expansion, etc.). If there is a vibration problem, this transient behavior must be improved with a balance weight, which is a non-transient, constant countermeasure.

The balancing process therefore often can only be a compromise (improving the base load condition but making it worse at idle). The balanceable part or component of the vibration signature (1X component) is what we call the unbalance. This is only one vibration cause among many others, but in most cases an essential contribution to the vibration signature.

The dynamic properties of the bearings in which the rotor is suspended are always different from each other, which makes the balancing procedure more complex than in the factory on a balancing machine. To be able to say whether a given vibration is due to unbalance, and whether it can be corrected by rebalancing of the system, certain information is required. At first, the original vibration situation must be judged in order to make the decision if balancing should be carried out.

Measuring conventions:

The measured vibration data consisting of magnitude and phase information are normally plotted on a vector diagram. It has 0° horizontal right and 90° top vertical. The equivalent machine rotation sense is in direction of decreasing angels. To balance a machine, we measure the responses of shaft or bearings to the kinetic imbalance forces regarding amount and phase angle. This is done at or close to the supporting points. Therefore, we always use the same measuring conventions.

Figure 4.3 shows the phase angle in relation to the sense of rotation. It should be always used in the same way to allow cross-comparisons of different actions.

Fig. 4.3 Polar plot for drawing the measuring values

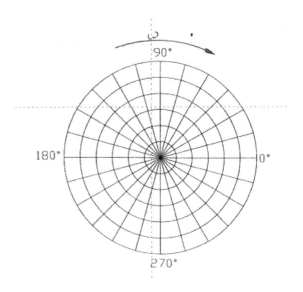

A few words about the phase angle (see Fig. 4.4).

Fig. 4.4 The phase angle

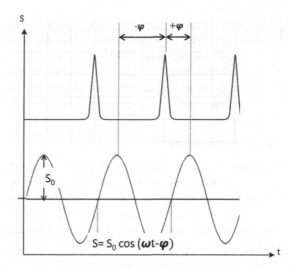

The phase angle is the time elapsed between the passage of the "0" mark and the next vibration maximum. The "0"-mark starts a (virtual) stopwatch and the vibration maximum stops it again. Here, one time period of the revolution will be expressed in angular units from 0° to 360°. At the machine, the reference transducer or key phasor is in the most cases top vertical. If an additional optical sensor has to be attached, it should be top vertical. It makes balancing easier, when the phase sensor (key phasor) is in line with the vibration transducers. Either the key-phasor mark or the attached optical tape defines 0°. It must be at the 0° mark of the shaft. 90° will come after the 0° mark. That means that the degree and/or the balancing hole numeration is decreasing regarding the sense of rotation. An outside observer, however, will see increasing numbers/angles through the balancing hole when the machine is on baring gear. Figure 4.5 shows the measurement conventions regarding the rotation sense and the angle numeration on the shaft.

Fig. 4.5 Measuring conventions

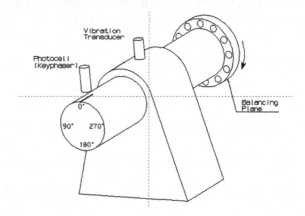

Measurements before balancing, reference measurements:

1. Cold start
2. Idle until vibrations are stable but at least one hour
3. Loading to base load/load under normal operation
4. Wait until vibrations are stable
5. De-loading to idle
6. Idle until vibrations are stable but at least one hour
7. Warm run-down
8. Warm start.

Balancing procedure:

For field balancing, we use the "influence vector" or "influence coefficient" method. That means that in every balance plane of the rotor, we determine experimentally an influence vector of a balance weight (as shown above in Fig. 4.2).

As soon as we have this information for all balance planes, the obtained influence vectors are optimized in order to obtain minimum vibrations at the bearings. This optimization process can be done with a mathematical algorithm in a computer (balancing program) or graphically by hand. We made the experience that influence vectors measured at a certain power plant are useable at another jobsite, as long as we work on the same machine type. Therefore, an influence vector database has been established for the different machine types.

There is one fundamental rule in jobsite balancing, which must be always considered:

Balancing is a process, where we always repeat runs under the same operating conditions such as:

- first run: reference run without weight,
- second run: first trial weight or weight set,
- third run: second trial weight or weight set and
- fourth run: final run with balance weight.

Like with this electric motor shown in Fig. 4.6, the same minimum main measuring positions (at least at every bearing) and measurement quantities apply also for a gas or steam turbine. Primarily, the relative shaft vibration will be measured, but the bearing vibrations as well.

Fig. 4.6 Minimum
measuring positions (Brüel
and Kjäer Vibro 1995)

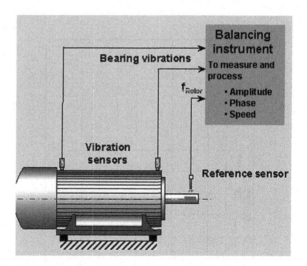

Fig. 4.6 Minimum measuring positions (Brüel and Kjäer Vibro 1995)

Before starting the balancing exercise, we prepare two forms:

i. one table for the vibration readings and
ii. one polar graph for plotting the measured values.

The vibration readings on which the balance exercise is based are preferably (relative) shaft vibrations and if these are not available vertical pedestal vibrations. Figures 4.7 and 4.8 are examples for these forms (Table 4.2).

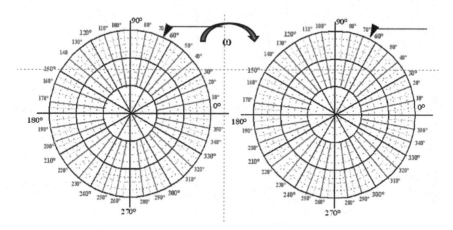

Fig. 4.7 Polar plot forms for plotting measured vibration values

Table 4.2 Sample for the vibration readings table

Vibration Readings

Date Time	Run Nr.	Ref Nr.	n U/min	Power MW	VE Amp	VE ∠	HO Amp	HO ∠	RI Amp	RI ∠	LE Amp	LE ∠	VE Amp	VE ∠	HO Amp	HO ∠	RI Amp	RI ∠	LE Amp	LE ∠	Bal. Weights plane	Bal. Weights mass g	Bal. Weights no./a	Code	Remarks

Experience has shown that the rotors, which form the shaft train in a plant, are very well mechanically balanced in the factory, since computer-based influence vector methods are used. But there can be various reasons for rebalancing as well as various kinds of unbalance causes at jobsite operation of a gas or steam turbine plant (but mostly thermal unbalances, because of changes of the rotors, when exposed to temperature).

Therefore, also vibration readings for all kinds of operating conditions are needed as an information source for balancing. Throughout our experience in jobsite balancing, the following standard program has been established:

From the operating conditions mentioned above, the one must be selected in which the vibrations must primarily be reduced. This is our reference run and the basis for the weight influence vectors. But the vibrations at the other conditions must also be considered.

A suitable operating condition, at which the balancing exercise will be carried out, should be a condition at which we obtain stable temperature resp. vibration conditions. This is preferably base load. It will often take a few hours, especially after a cold start, to get a stable vibration trend!

As soon as the vibrations are stable, the first reading, without any balance weight, is taken. We call it the "zero-run" or "reference run."

An example of a balancing exercise is given below. We assume the following values for our reference run:

- 4.5 mm/s at the turbine bearing, phase: 61° and
- 5.1 mm/s at the compressor bearing, phase: 100°.

These values are drawn into the polar plot and marked with "0" to indicate the reference run, and thus the vibration readings before balancing (see Fig. 4.8).

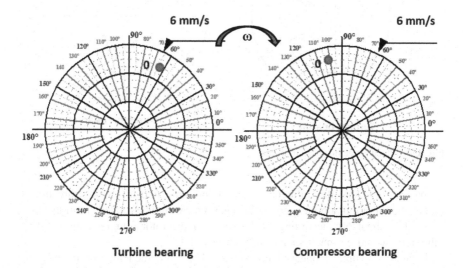

Fig. 4.8 Reference run measurements in the polar plot

Now the machine is stopped to attach the first trial weight. When the shaft is standing, we need to switch off the baring gear in order to attach the weight. The circumference position and the standing time must be recorded.

After the weight has been attached (by means of the proper balancing tools), the shaft must be rotated 180° plus the original standing position. Then the same time must be waited, before engaging the baring gear again. A hot turbine can only be stopped for maximum of 10 min!

As soon as the weight is attached, we put the machine at the identical operating condition as at the reference run, wait for stabilization of the vibration trend and make a new reading.

We chose as a weight of 1000 g at 0° circumference position in the balancing plane of the compressor. The new readings with the weight attached are:

- 3.9 mm/s at the turbine bearing, phase: 350° and
- 4.1 mm/s at the compressor bearing, phase: 55°.

These new vibration readings are again drawn into the same polar plot as the reference run and marked with "1" to indicate the first trial run (see Fig. 4.9). When we connect "0" with "1" with an arrow, this arrow represents the influence vector of the weight. Now we have the influence vector 0–1 for the trial run.

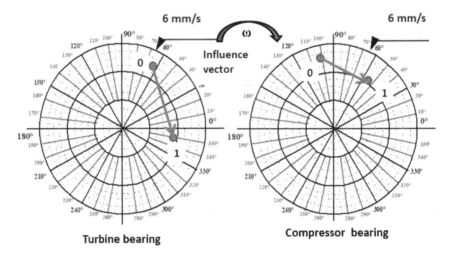

Fig. 4.9 Influence vector of the trial run

We see now that this influence vector is pointing into the wrong direction to reduce the vibration at both bearings. If this vector would be rotated (from its origin point "0") about 60° according to the sense of the rotor rotation , the resulting vibrations would be much smaller at both bearings.

Balancing assumes a linear kinetic system. Therefore, the length of the influence vector is directly proportional to the amount of the balancing weight and the direction of the vector represents the circumference position of the weight in the balancing plane. If we want to move the vector by 60° according to the sense of machine rotation, we must move our trial weight by approx. 60° according to the sense of machine rotation. If the vector should become longer, we must increase the trial weight proportionally. So, we take out the weight of the compressor plane at position 0° and plug it in the same plane at 300°. This will be the run #2. After the machine has been started, brought to our reference operating condition, we will read the following vibration value after stabilization:

- 1.2 mm/s at the turbine bearing, phase: 170° and
- 1.3 mm/s at the compressor bearing, phase: 92°

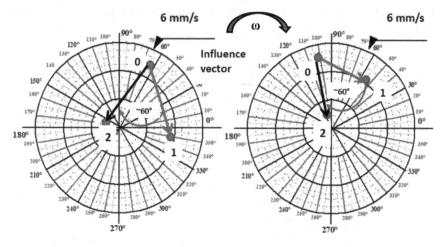

Fig. 4.10 Indicates the final run 2

The trial run will be optimized to obtain the vector #2 according to Fig. 4.10.

For a balancing exercise from scratch at a flexible rotor, three fundamental questions must be considered:

- Which is the most effective balancing plane?
- Where is the correct circumference position at this plane?
- What will be the size of the balance weight?

At a balance exercise, you normally plot all the important measuring positions (the 4 shaft vibration positions on 4 polar plots) simultaneous and you work through the balance planes plane by plane sequentially.

You normally will achieve an antiphase influence vector at the two end-bearings with a weight in the end-planes, turbine or compressor. A center-plane weight normally causes an in-phase influence vector at the end-bearings.

When you start a balancing exercise, you will use already taken influence vectors from other plants where a machine had been balanced of the same type. If you need start newly from scratch, you do not want to place your test weight in a completely wrong plane or position. Following are some hints to determine the effective balance plane resp. a circumference position, which at least does not aggravate the current situation:

Balance plane:

In most of the cases, the unbalance sits close to the bearing with the highest (relative) shaft vibration. But there are cases where this consideration is misleading (e.g., critical speeds or structural resonances close to the balancing speed). Then we can take advantage of the following typical kinetic behavior of oscillating systems. In Fig. 4.11, we see how to conclude to the optimal balance plane from the polar plots.

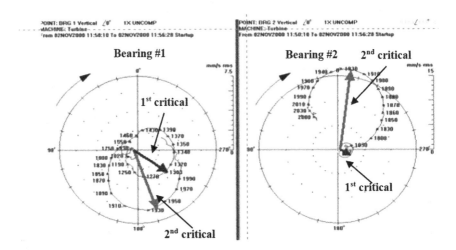

Fig. 4.11 Balance plane determination

Figure 4.11 shows two polar plots—the left plot of bearing #1 and the right plot of bearing #2. The 1st critical is marked with a blue arrow and the 2nd critical with a red arrow. Unfortunately, the vibrations at bearing #2 are small, but still the phase angle can be recognized.

A mode shape (kinetic deformation at the critical speed) will hardly change (or only show a minimum change of) its phase at the excitation point, when it changes from a lower to a higher mode (or critical speed).

According to the example of Fig. 4.11, the phase angles (preferably on a polar plot of the shaft vibrations) must be watched when the rotor passes the first and second critical speed. In our example, we see a much smaller phase difference between 1st (blue arrow) and 2nd critical (red arrow). Therefore, we can state that the predominant unbalance share is closer to bearing #1. We will start at the turbine balancing plane close to bearing #1, despite we have lower vibrations at the turbine bearing #2.

The information about the circumference position of the unbalance is contained in the phase angle. The instrument indicates the high point of motion regarding the reference (key phasor) mark. The high point of motion corresponds to the heavy spot, but only for rigidly behaving rotors, which are rotors having their first resonance (critical speed) higher than the operating speed. Our turbine rotors must pass normally two critical speeds until rated speed is reached. These rotors behave elastically. A statement about the heavy spot can only be made in the "rigid range," which is below the first critical of the rotor (or rotor section) because there are no disturbing phase changes (see Fig. 4.12).

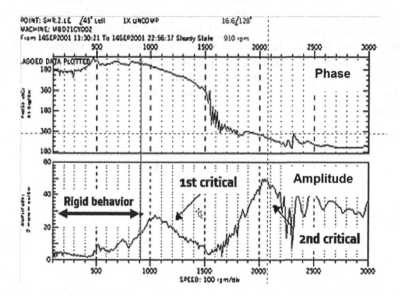

Fig. 4.12 Rigid speed range below the 1st critical

The fundamental aim of balancing is to find the heavy spot and place a compensation weight at the opposite side on the circumference. To determine the heavy spot at rated speed is in most cases misleading because of the flexible nature of rotor and support system (this had been proven by experience due to the various phase changes caused by the resonances appearing during the speed change).

Figure 4.12 shows a run-up Bode plot of a 250 MW gas turbine. The first critical we see is at a speed of approx. 1050 rpm.

Looking at the phase angle plot (upper graph of Fig. 4.12), we see an almost constant phase angle until close to the first critical. The speed range from standstill upward represents the non-flexible or rigid range of the rotor before it enters the first critical speed and becomes flexible.

In this speed range, the vibration reaction to exciting forces is only dependent on the spring properties of the rotor and support. Inertia and/or damping forces will cause no time (-phase) differences between driving force and reaction. Therefore, there are no phase changes between excitation (unbalance) and response (vibration).

Looking at the equations of motion (see Eqs. (1.23) and (1.24) or (1.27) and (1.28) in Chap. 1.3), only the last term for the spring restoring force $(cr_y$ resp. $cr_z)$ will be actively controlling the response. The first two terms (inertia and damping forces) will become active as soon as the rotors become flexible.

We can take the phase angle as the circumference position of the heavy spot and put the balance weight on the opposite side. In this case, it is approx. 180°, and we put the weight a 0°.

Figure 4.13 shows the polar plot of a run-up of a GT8. The first critical speed is clearly visible as a loop (with a changing direction—increasing phase angels from low to higher speed always against the rotation sense). This loop represents the flexible behavior passing the critical (high point at 5230 rpm).

But first we look at the speed from standstill, where the vibration response increases in amplitude, but no phase change appears (0–4930 rpm). This is the speed range where spring restoring forces control the vibration response.

The heavy spot of a rotor will be determined from the phase angle of the rigid speed range (see Fig. 4.13).

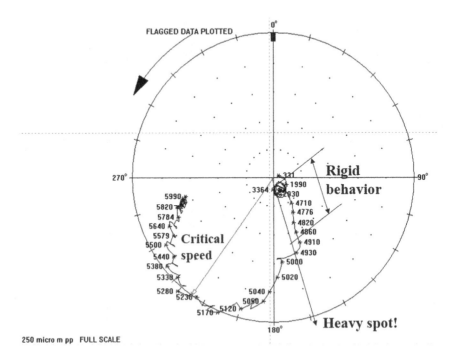

Fig. 4.13 Heavy spot and the 1st critical

Size of the weight:

The size of a first trial weight is largely dependent upon experience. It must have a certain amount, not too small to override the disturbing influences and not too big to allow the machine reaching rated speed.

For a 250–350 MW gas turbine or the LP of a 200 MW steam turbine, it is recommended starting with 600–700 g for the end planes. For the center plane, 300–400 g are recommended. The test weights for other rotors must be adjusted according to rotor speed, rotor weight and balancing diameter. The above-mentioned recommendations apply for 50 Hz–3000 rpm machines.

Important rules of balancing:

- Take readings for weight influences only at stable vibration conditions.
- Start your reference (trial, balance run or datum run) always with the same initial situation (same rotor temperature).
- Maintain always the same operating condition and the same sequence in changing operating conditions and taking vibration readings (taking readings at a certain part load coming from idle and comparing them with readings coming from base load is not allowed!)
- Be always aware that the vibration changes you will register should solely be due to the weight changes and to nothing else.

Chapter 5
Vibration Limits

Here are the vibration limits from ISO ((International Organization for Standardization). They had been issued by the ISO between 1995 and 2009 by the vibration committee. The authors of this book had also been members of this committee for many years and had also been precipitant for the commitment of these Standards.

Practical experience proofed that these Standards protect the machines sufficiently at one hand. At the other hand, they allow enough margin to pinpoint the problem. The vibration limits according to ISO 10816 (bearing vibrations) and ISO 7919 (shaft vibrations) are as follows:

Classification of vibration evaluation zones:
The following evaluation zones are defined to permit a qualitative assessment of the shaft vibration of a given machine and to provide guidelines on possible actions.

- Zone A: The vibration of newly commissioned machines would normally fall within this zone.
- Zone B: Machines with vibration within this zone are normally considered acceptable for unrestricted long-term operation.
- Zone C: Machines with vibration within this zone are normally considered unsatisfactory for long-term continuous operation. Generally, the machine may be operated for a limited period in this condition until a suitable opportunity arises for remedial action.
- Zone D: Vibration values within this zone are normally considered to be of sufficient severity to cause damage to the machine.

NOTE: The evaluation zones defined above are relevant to normal steady-state operation at rated speed.

© Springer Nature Switzerland AG 2020
F. Herz and R. Nordmann, *Vibrations of Power Plant Machines*,
https://doi.org/10.1007/978-3-030-37344-3_5

Classification of relative shaft vibrations or large gas and steam turbo-sets during run-up, run-down, over-speed:

- For speeds greater than 0.9 times the normal operating speed, the vibration magnitude corresponding to the zone boundary C/D for the maximum normal operating speed.
- For speeds less than 0.9 times the normal operating speed, 1.5 times the vibration magnitude corresponding to the zone boundary C/D for the maximum normal operating speed.

Classification of bearing vibrations for large steam turbo-sets:
See Table 5.1.

Table 5.1 Bearing vibration limits for large steam turbo-set (International Organisation for Standardization 1996)

Zone boundary	Shaft rotational speed (rpm)	
	1500 or 1800	3000 or 3600
	Vibration velocity [mm/s (r.m.s.)]	
A/B	2.8	3.8
B/C	5.3	7.5
C/D	8.5	1.8

Note These values relate to steady-state operating conditions at rated speed for the recommended measurement locations. They apply to radial vibration measurements on all bearings and to axial vibration on thrust bearings

Classification of bearing vibrations for large gas turbo-sets:
See Table 5.2.

Table 5.2 Bearing vibration limits for large gas turbo-sets (International Organisation for Standardization 1998)

Zone boundary	Vibration velocity [mm/s (r.m.s.)]	
A/B	2.8	3.8
B/C	5.3	7.5
C/D	8.5	1.8

Note These values, which are the upper limits of zones A, B and C, respectively, should apply to radial vibration measurements on all bearing housings or pedestals and to axial vibration measurements on housings containing an axial thrust bearing, under steady-state operating conditions at rated speed

Classification of shaft vibrations for large steam turbo-sets:
See Table 5.3.

Table 5.3 Shaft vibration limits of large steam turbo-sets (International Organisation for Standardization 1996)

Zone boundary	Shaft rotational frequency (rpm)				
	1500	1800	3000	3600	
	Maximum absolute displacement of shaft (μm p.p.)				
A/B	120	110	100	90	
B/C	240	220	200	180	
C/D	385	350	320	290	
Zone boundary	**Shaft rotational frequency (rpm)**				
	1500	1800	3000	3600	
	Maximum relative displacement of shaft (μm p.p.)				
A/B	100	90	80	75	
B/C	200	185	165	150	
C/D	320	290	260	240	

Classification of relative shaft vibrations for large gas turbo-sets:
    See Fig. 5.1.

Fig. 5.1 Relative shaft
vibration limits of large gas
turbo-sets

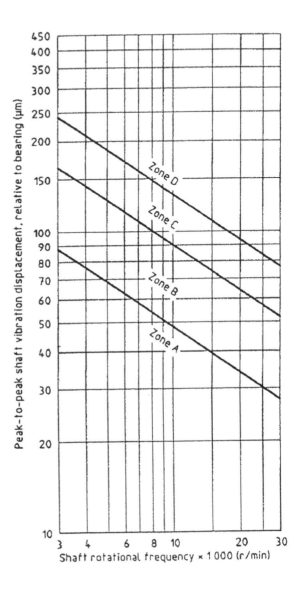

Chapter 6
Some Hints to Identify Vibration Problems

This is a list of practical hints in order to identify vibration problems. They can direct the solution process into the right direction and could be the basis of the problem elimination. All these hints are coming from practical experience derived from the case studies in Chap. 3 and are the product of a 50-year experience.

- Be aware of the dominating vibration frequency whether it is rotational 1X or its harmonics 2X, 3X or subharmonic below 1X like 0.45X or a critical speed frequency.
- In case of 1X (rotational or unbalance frequency), look at the relationship between shaft and pedestal vibrations: high shaft vibrations can mean any kind of unbalance.
- High pedestal vibration can mean structural resonance problems. Another feature of those resonances is the fact that there is a pronounced direction of dominant vibrations (axial, horizontal or vertical).
- In case of structural resonances find the spot of the biggest relative motion between elements (pedestal-foundation, etc.), this is usually the weak spot.
- 2X vibrations are mainly created by the generators, either by:
  - Sag excitation by the rotor: the vibration is not dependent on operation parameters.
  - Magnetic excitation by the stator: the vibration will appear as soon as the field current is engaged.
- Newly appearing 2X vibrations, especially at run-down and half speed of the 1st critical, might indicate a lateral rotor crack.
- An increase in vibration at a warm run-down compared to a cold run-up as well as a higher pronounced 1st critical usually indicates a thermal unbalance of the rotor.
- Vibrations often react, when the load changes. This can be due to temperature changes or the torque changes, when load changes. Vibrations lag behind the temperature changes, but they react immediate together with the torque change.

© Springer Nature Switzerland AG 2020
F. Herz and R. Nordmann, *Vibrations of Power Plant Machines*,
https://doi.org/10.1007/978-3-030-37344-3_6

References

Brüel, & Kjäer Vibro. (1995). *Measurement types and sensors for machine diagnosis.*
Brüel, & Kjäer Vibro. (1995). *Multi-plane balancing.*
Brüel, & Kjäer Vibro. (2002). *Analysis of the dynamic behavior of machines.*
Gasch, R., & Pfützner, H. (1975). *Rotordynamik.* Heidelberg, New York: Springer.
Harris, C. M. (1961). *Shock and vibration handbook.* New York: McGraw-Hill Publishing Company.
International Organisation for Standardization. (1996, February). ISO 10816-2:1996. *Mechanical vibration—Evaluation of machine vibration by measurements on non-rotating parts – Part 2: Large land-based steam turbine generator sets in excess of 50 MW.* International Organisation for Standardization.
International Organisation for Standardization. (1996, July). ISO 7919-2:1996. *Mechanical vibration of non-reciprocating machines—Measurements on rotating shafts and evaluation criteria—Part 2: Large land-based steam turbine generator sets.* International Organisation for Standardization.
International Organisation for Standardization. (1996, July). ISO 7919-4:1996. *Mechanical vibration of non-reciprocating machines—Measurements on rotating shafts and evaluation criteria—Part 4: Gas turbine sets.* International Organisation for Standardization.
International Organisation for Standardization. (1998, July). ISO 10816-4:1998. *Mechanical vibration—Evaluation of machine vibration by measurements on non-rotating parts—Part 4: Gas turbine driven sets excluding aircraft derivatives.* International Organisation for Standardization.
International Organization for Standardization. (1995, December). ISO 10816-1:1995. *Mechanical vibration—Evaluation of machine vibration by measurements on non-rotating parts.* International Organization for Standardization.
International Organization for Standardization. (1996, July). ISO 7919-1:1996. *Mechanical vibration of non-reciprocating machines—Measurements on rotating shafts and evaluation criteria—Part 1: General guidelines.* International Organization for Standardization.
Protectiv Supplies & Procurement Services. (2019, 15 May). *Bently Nevada ADRE 408 DSPi portable condition monitoring system.* Retrieved from https://www.protectivesupplies.com/item/Brand_BentlyNevadaADRE408DSPiPortableConditionMonitoringSystem_483_0_4250_1.html.

Printed in the United States
by Baker & Taylor Publisher Services